Lecture Notes in Statistics

125

Edited by P. Bickel, P. Diggle, S. Fienberg, K. Krickeberg, I. Olkin, N. Wermuth, S. Zeger

Valerii V. Fedorov
Peter Hackl

Model-Oriented Design of Experiments

Springer

Valerii V. Fedorov
Oak Ridge National Laboratory
P.O. Box 2008
Oak Ridge, TN 37831-0608
USA

Peter Hackl
University of Economics and Business Administration
Department of Statistics
Augasse 2
A 1090 Wien, Austria

Fedorov. V. V. (Valerii Vadimovich)
  Model-oriented design of experiments/Valerii V. Fedorov, Peter
Hackl.
    p.   cm.—(Lecture notes in statistics; 125)
  Includes bibliographical references and index.
    ISBN 0-387-98215-9 (softcover: alk. paper)
    1. Experimental design I. Hackl, Peter. II. Title.
    III. Series: Lecture notes in statistics (Springer-Verlag); v. 125.
  QA279.F43 1997
  519.5—dc21   97-15703

Printed on acid-free paper.

Camera-ready copy provided by the author.s.
Printed and bound by Braun-Brumfield, Ann Arbor, MI.
Printed in the United States of America.

9 8 7 6 5 4 3 2 1

ISBN 0-387-98215-9   Springer-Verlag New York Berlin Heidelberg   SPIN 10523539

Preface

These lecture notes are based on the theory of experimental design for courses given by Valerii Fedorov at a number of places, most recently at the University of Minnesota, the Vienna of University, and the University of Economics and Business Administration in Vienna.

It was Peter Hackl's idea to publish these lecture notes and he took the lead in preparing and developing the text. The work continued longer than we expected, and we realized that a few thousand miles distance remains a serious hurdle even in the age of Internet and many electronic gadgets.

While we mainly target graduate students in statistics, the book demands only a moderate background in calculus, matrix algebra and statistics. These are, to our knowledge, provided by almost any school in business and economics, natural sciences, or engineering. Therefore, we hope that the material may be easily understood by a relatively broad readership.

The book does not try to teach recipes for the construction of experimental designs. It rather aims at creating some understanding – and interest – in the problems and basic ideas of the theory of experimental design. Over the years, quite a number of books have been published on that subject with a varying degree of specialization. This book is organized in four chapters that lay out in a rather compact form all ingredients of experimental designs: models, optimization criteria, algorithms, constrained optimization. The last third of the volume covers topics that are relatively new and rarely discussed in form of a book: designs for inference in nonlinear models, in models with random parameters, in stochastic processes, and in functional spaces; for model discrimination, and for incorrectly specified (contaminated) models.

Data collected by performing an experiment are based on two elements: (i) a clearly defined objective and (ii) a piece of real world that generates – under control of the experimenter – the data. These elements have analogues in the statistical theory: (i) the optimality criterion to be applied has to be chosen so that it reflects appropriately the objective of the experimenter, and (ii) the model has to picture – in adequate accuracy – the data generating process.

When applying the theory of experimental design, it is perhaps more true than for many other areas of applied statistics that the complexity of the real world and the ongoing processes can hardly be adequately captured by the concepts and methods provided by the statistical theory. This theory contains a set of strong and beautiful results, but it permits in only rare cases closed-form solutions, and only in special situations is it possible to construct unique and clear-cut designs for an experiment. Planning an experiment means rather to work out several scenarios which together yield insights into and understanding of the data generating process, thereby strengthening the intuition of the experimenter. In that sense, a real life experiment is a compromise between results from statistical theory and the *a priori* knowledge and intuition of the experimenter.

We have kept the list of references as short as possible; it contains only easy accessible material. We hope that the collection of monographs given in References will be sufficient for readers who are interested in the origin and history of the particular results. A bibliography related to the more recent results can be found in the papers by Cook and Fedorov (1995) and Fedorov (1996). Note that Volume 13 of the Handbook of Statistics, edited by Ghosh and Rao (1996), consists entirely of survey-type papers related to experimental design.

We gratefully acknowledge the help and encouragement of friends and colleagues during the preparation of the text. Debby Flanagan, Grace Montepiedra, and Chris Nachtsheim participated in the development of some results from Chapter 5; we are very grateful for their contributions. We are thankful to Agnes Herzberg, Darryl Downing, Max Morris, Werner Müller, and Bob Wheeler for discussions and critical reading of various parts of this book. Stelmo Poteet and Christa Hackl helped us tremendously in the preparation of the text for publication.

Valerii V. Fedorov, Oak Ridge
Peter Hackl, Vienna
December 1996

Contents

Introduction

The collection of data requires a certain amount of effort such as time, financial means, or other material resources. A proper design potentially allows to make use of the resources in the most efficient way.

The history of publications and the corresponding statistical theory goes back as far as to 1918 when Smith (1918) published a paper that presents optimal designs for univariate polynomials up to the sixth degree. However, the need to optimize experiments under certain conditions was understood by many even earlier. Stigler (1974) provides an interesting historical survey on this subject. After some singular earlier work, the core of theory of optimal experimental design was developed during the fifties and sixties. The main contribution done during that time is due to Jack Kiefer. A survey of Kiefer's contribution to the theory of optimal design is contained in the paper by Wynn (1984). Brown et al. (1985) published Kiefer's collected papers. Other important names and papers from that early times may be found in Karlin and Studden (1966), Fedorov and Malyutov (1974), and Atkinson and Fedorov (1988). Box and coauthors discussed related problems associated with actual applications; see, e.g., Box and Draper (1987). The work of the Russian statisticians that covers both mathematical theory and algorithms is surveyed by Nalimov et al. (1985).

The first comprehensive volume on the theory of optimal experimental design was written by Fedorov (1972). The book by Silvey (1980) gives a very compact description of the theory of optimal design for estimation in linear models. Other systematic monographes were published by Bandemer et al. (1977), Ermakov (1983), Pázman (1986), Pilz (1993), and Pukelsheim (1993). A helpful introductory textbook is Atkinson and Donev (1992).

Models and Optimization Problems. In the description of experiments we distinguish between

- variables that are the focus of our interest and response to the experimental situation, and

- variables that state the conditions or design under which the response is obtained.

The former variables usually are denoted by y, often indexed or otherwise supplemented with information about the experimental conditions. For the latter we distinguish between variables x that are controlled by the experimenter, and variables

1

t that are – like time or temperature in some experiments – out of the control of the experimenter. In real-life experiments, y is often and x and t are almost always vectors of variables. The theory of optimal designs discussed in this book is mainly related to the linear regression. But various extensions comprise

- multi-response linear regression,

- nonlinear regression,

- regression models with random parameters,

- models that represent random processes,

and other generalizations of the regression model concept including discrimination between competing models.

The set of values at which it is possible and intended to observe the response is called the design region X. In general, X is a finite set with dimension corresponding to the number of design variables. The classical design theory has been derived for this case. However, in real-life problems we often encounter design restrictions. In a time-series context, it is typically not possible to have multiple observations at the same time point, so that the design region consists of an (equidistant) grid in time. Similar restrictions might be requested due to geographical conditions, mixing constraints, etc. The most common cause for restrictions are due to cost limitations: costs often depend on the design point; e.g., the investment and maintenance costs of a sensor can strongly be determined by the accessibility of its location. A crucial but purely technical reason for a restriction that has to be taken into account in all practical applications is the fact that the distribution of the observations in the design region must be discrete.

The classical optimal design problem is the estimation of the model parameters subject to the condition that a *design criterion* is optimized. In the case of a simple linear regression with $E\{y\} = \beta_0 + \beta_1 x$, the variance of the estimate β_1 is proportional to the reciprocal of the mean squared deviations between the design points x_i, and their mean $\bar{x}$: $\text{Var}\{\hat{\beta}_1\} \propto 1/\sum_{i=1}^{N}(x_i - \bar{x})^2$. Consequently, we can – given the number N of observations – minimize the variance of the estimate by shifting one half of the design points to each of the limits of the design region.

In cases where the parameter vector has $k > 1$ components, a possible design criterion is a (scalar) function of the corresponding covariance matrix. As it is intuitively obvious, the choice of the design criterion will turn out to be a crucial part of an optimal design problem, both from the aspect of technical effort and interpretation of results. The design problems discussed in the literature go beyond the estimation of model parameters. Design criteria similar to those used in estimating model parameters can be applied for

- estimating functions of the model parameters,

- constructing a statistical test with optimal power, or

- screening for significant variables,

- discriminating between several models.

Illustrating Examples

In many practical cases y, x and t are vectors in the Euclidean space. For instance, y may be the yield of crop(s), x are concentrations of fertilizers, and t are weather conditions. Experimental data let us infer – to the desired degree of accuracy and reliability – what dosage of an fertilizer is optimal under certain conditions. In many cases it will help the reader to achieve a better understanding of the general theoretical results (Chapters 1 to 4) if she or he tries to relate these with this or a similar situation. The recent development shows (see Chapter 5) that the main ideas of the optimal experimental design may be applied to the large number of problems, in which y, x, and t have more complicated structures. We sketch a few examples that are typical for various situations where design considerations can be used to economize the experimental effort in one or another way.

Air Pollution

The air pollution that is observed in a certain area is determined, among other factors, by the time of observation, by the location of the sensor, and by the direction of wind. The wind direction determines what sources of immission are effective at the location of observation. The air pollution is measured in terms of the concentration y of one or several pollutants; the location of the observation station is $x = (x_1, x_2)^T$, the wind direction is described by $v = (v_1, v_2)^T$, and time is denoted by t. Typical sets of information are:

1. the function $y(x, v, t)$, $0 \leq t \leq T$, for a given location x;

2. the vector $[y_1(x, v, t), \ldots, y_k(x, v, t)]^T$ for a given location x at time t; k is the number of pollutants;

3. the scalar
$$y(x) = T^{-1} \int_v \int_0^T y(x, v, t) \, p(v) \, dt dv \,,$$

   i.e., the mean air pollution at location x in the period $[0, T]$ and with respect to distribution $p(v)$ of the wind direction.

The design problem in this context could be: Where should we locate sensors so that the result of a certain analysis has maximal accuracy? Sections 2.6, 5.1, and 5.3 may help to answer such questions.

ph-Profiles

If plaque on the tooth surface is exposed to carbonhydrate the plaque pH rapidly decreases from the neutral pH value of about 7.0 to a minimum and later increases slowly to approach the neutral status. This pH-profile can be modelled as

$$y(t) = \beta[1 + e^{-\theta_1 t} - e^{-\theta_2 t}],$$

where the parameters β, θ_1, and θ_2 represent the neutral pH value, the decrease rate and the increase rate, respectively. A standard design for measuring the pH-profile or some characteristic of it is to take equally spaced observations over a certain time interval. Can we find a pattern of measurements that give maximal precision of the estimates at a given number of observations or at given costs? The corresponding type of model is considered in Section 5.2.

Chemical Reactor

The output y of a chemical process that is going on in a reactor is determined by the amounts x_1 and x_2 of two input components, and by the profile $x_3(s)$ of the temperature over the length s of the reactor. The design problem in this context could be: Find the set $\{x_1, x_2, x_3(s)\}$ that provides an optimal approximation of y. Note that $x_3(s)$ is an element in some functional space. A suitable design technique can be found in Section 5.1; although this section refers to time-dependence, the discussed methods may be applied to a broader class of problems.

Spectroscopic Analysis

To describe a compound spectrum, knowledge of the concentration θ_i of the spectral components $f_i(\nu)$, $i = 1, \ldots, m$, is needed for certain frequencies ν. These components can be assessed by observing the spectrum in certain "windows", i.e., subsets of the entire frequency domain. Such a window is defined by an indicator function $x_j(\nu)$ which has the value one for the frequency interval of interest and zero elsewhere. The observed signal for the given window $x_j(\nu)$ is

$$y(x) = \int \sum_{i=1}^{m} \theta_i \, f_i(\nu) \, x_j(\nu) \, d\nu.$$

The design problem is to choose windows $x_j(\nu)$, $j = 1, \ldots, n$, in such a way that the best estimation of all or selected θ_i is possible.

The Book's Outline

The first four chapters cover general material. In particular, Chapter 1 contains a very short collection of facts from regression analysis. Chapter 2 is essential for understanding and describes the basic ideas of convex design theory. The subsequent chapters concern the numerical methods (Chapter 3) and a few theoretical extensions (Chapter 4). The reader may abstain from detailed reading of the sections on numerical methods; some basic algorithms are already available either in widely used statistical packages, e.g., SAS, JMP, and SYSTAT, or in more specific software like ECHIP and STAT-EASE. Chapter 5 is the largest chapter and describes applications of the convex design theory to various specific models. The appendix includes a rather standard collection of formulas (mainly from matrix algebra) and is included for the reader's convenience.

Chapter 1

Some Facts From Regression Analysis

1.1 The Linear Model

Almost all concepts and methods to be discussed in this book are related to the linear regression model

$$y = \eta(x, \theta) + \varepsilon = \theta^T f(x) + \varepsilon. \tag{1.1.1}$$

The variable y is called the response (or dependent or observed) variable, and mostly we assume that it is scalar. More complicated structures will occasionally be considered.

The error ε is assumed to be random. If not stated otherwise, we assume

$$E\{\varepsilon\} = 0 \text{ and } E\{\varepsilon^2\} = \sigma^2 \tag{1.1.2}$$

or, as an immediate result,

$$E\{y|x\} = \theta^T f(x) \text{ and } \text{Var}\{y|x\} = \sigma^2; \tag{1.1.3}$$

E and Var stand for expectation and variance, respectively. For observations y and y' with errors ε and ε', respectively,

$$E\{\varepsilon \varepsilon'\} = 0 \text{ and } \text{Cov}\{y, y'|x\} = 0, \tag{1.1.4}$$

where Cov stands for the covariance. Let us note that the assumptions concerning the randomness of ε are crucial. While many results stay valid for errors of a different nature (e.g., errors based on a deterministic approach), most of them are formulated in terms of probability theory and it may take some effort to interpret them in another setting.

The response function $\eta(x, \theta)$ is linear in the components of the vector

$$f^T(x) = (f_1(x), \dots, f_m(x))$$

7

that are called "basis" functions and assumed to be known. The variables

$$x^T = (x_1, \ldots, x_k)$$

are called predictors or independent or control or design variables. The parameters $\theta^T = (\theta_1, \ldots, \theta_m)$ are typically unknown; methods of inference for these parameters are the main interest of the book.

Whenever it is not confusing the same symbol will be used for a random variable and its observed value.

The Least Squares Estimator. Let r_i, $i = 1, \ldots, n$, be the number of observations y_{ij}, $j = 1, \ldots, r_i$, that were made at the design or support point x_i. The collection of variables

$$\xi_N = \left\{ \begin{array}{c} x_1, \ldots, x_n \\ p_1, \ldots, p_n \end{array} \right\} = \left\{ \begin{array}{c} x_i \\ p_i \end{array} \right\}_1^n \qquad (1.1.5)$$

with $p_i = r_i/N$ and $N = \sum_{i=1}^n r_i$ is called the design (of the experiment).

Based on the observations

$$y_{11}, \ldots, y_{1r_1}, \ldots, y_{n1}, \ldots, y_{nr_n}$$

the least squares estimator (l.s.e.) $\hat{\theta}$ for θ is defined to be

$$\hat{\theta} = \arg\min_\theta v(\theta), \qquad (1.1.6)$$

where

$$v(\theta) = \sum_{i=1}^n \sum_{j=1}^{r_i} \left(y_{ij} - \theta^T f(x_i) \right)^2. \qquad (1.1.7)$$

The l.s.e. will be of main interest in the following. Obviously, it can be written as

$$\hat{\theta} = \arg\min_\theta \sum_{i=1}^n p_i \left(\bar{y}_i - \theta^T f(x_i) \right)^2, \qquad (1.1.8)$$

where

$$\bar{y}_i = r_i^{-1} \sum_{j=1}^{r_i} y_{ij}.$$

It is well-known that the l.s.e. coincides with the best linear unbiased estimator (b.l.u.e.) [cf. the Gauss-Markov Theorem; see Rao (1973), Chpt. 4a]. Thus,

$$\mathrm{E}\{\hat{\theta}\} = \theta_t \text{ and } \mathrm{Var}\{\hat{\theta}\} \leq \mathrm{Var}\{\tilde{\theta}\}, \qquad (1.1.9)$$

where $\tilde{\theta}$ is any other linear unbiased estimator. The subscript t indicates the true values of the unknown parameters; it will be omitted when this does not lead to ambiguity, and this practice will be extended to other subscripts and superscripts whenever the interpretation remains clear. The ordering of positive (non-negative) definite matrices is understood as

$$A(\leq) < B, \text{ if } A = B + C, \ C > (\geq) 0,$$

where the latter inequality means that the matrix C is positive (non-negative) definite. The existence of the dispersion (variance-covariance) matrix $\mathrm{Var}\{\hat{\theta}\}$ implies that the design ξ_N is "good enough" for unique estimation of all components of the vector θ.

Setting the first derivative of the function v to zero gives the "normal equation"

$$\underline{M}(\xi_N)\,\hat{\theta} = Y\,, \tag{1.1.10}$$

where

$$\underline{M}(\xi_N) = \sigma^{-2} \sum_{i=1}^{n} r_i f(x_i) f^T(x_i) \tag{1.1.11}$$

and

$$Y = \sigma^{-2} \sum_{i=1}^{n} r_i\, \overline{y}_i f(x_i)\,. \tag{1.1.12}$$

It is obvious that $\hat{\theta}$ may uniquely be determined if the matrix $\underline{M}(\xi_N)$ is regular, i.e., if rank $\underline{M}(\xi_N) = m$, and the inverse matrix $\underline{M}^{-1}(\xi_N)$ exists. In this case

$$\hat{\theta} = \underline{M}^{-1}(\xi_N)\,Y\,. \tag{1.1.13}$$

Using the fact that

$$\mathrm{Var}\{LZ\} = L\,\mathrm{Var}\{Z\}\,L^T, \tag{1.1.14}$$

we find

$$\mathrm{Var}\{\hat{\theta}\} = \underline{M}^{-1}(\xi_N) = \underline{D}(\xi_N)\,. \tag{1.1.15}$$

Information Matrix. The matrix $\underline{M}(\xi_N)$ is usually called "information matrix". We want to emphasize that $\underline{M}(\xi_N)$ is determined by the design ξ_N but does not depend upon the observations y_{ij}. This means that the practitioner may select a design which is – in terms of $\mathrm{Var}\{\hat{\theta}\}$ – "better" than others. Actually, criteria for the quality of designs and procedures for selecting a "good" design are the main topics of the book.

Before going on we mention two important properties of the information matrix.

1. The information matrix is non-negative definite. The definition of a non-negative definite matrix implies its symmetry which follows for $\underline{M}(\xi_N)$ directly from definition (1.1.11). Non-negativeness, i.e., $a^T \underline{M}(\xi_N)\,a \geq 0$ for any vector $a \neq 0$, follows from the definition and the fact that $a^T f f^T a = (a^T f)^2 \geq 0$.

2. The information matrix is additive, i.e., it is the sum of information matrices that corresponds to the individual observations:

$$\underline{M}(\xi_N) = \sum_{i=1}^{n} r_i \underline{M}(x_i)\,, \tag{1.1.16}$$

where $\underline{M}(x) = \sigma^{-2} f(x) f^T(x)$. Obviously, adding an observation leads to an increase of the information matrix.

Singular Cases. If the information matrix is singular, i.e., rank $\underline{M}(\xi_N) = m' < m$, unique l.s.e.'s do not exist for all components of θ. However, a number of linear functions

$$\gamma = L\theta \tag{1.1.17}$$

may still be uniquely estimated if the solution

$$\hat{\gamma} = \arg\min_{\gamma = L\theta} \sum_{i=1}^{n} r_i \left(\overline{y}_i - \theta^T f(x_i)\right)^2 \tag{1.1.18}$$

is unique [see Rao (1973), Chpt. 4a]; the maximum number of components in γ is m'. For these functions (which are called "estimable under ξ_N") the estimator and its variance are

$$\hat{\gamma} = \underline{M}^-(\xi_N)Y \quad \text{and} \quad \text{Var}\{\hat{\gamma}\} = L\,\underline{M}^-(\xi_N)\,L^T, \tag{1.1.19}$$

where $\underline{M}^-$ stands for the pseudo- or g-inverse matrix of $\underline{M}$. Another way to determine $\hat{\gamma}$ and its variance is to use the regularized version of (1.1.13) and (1.1.14)

$$\hat{\gamma} = \lim_{\alpha \to 0} (\underline{M}(\xi_N) + \alpha P)^{-1}\,Y \tag{1.1.20}$$

and

$$\text{Var}\{\hat{\gamma}\} = \lim_{\alpha \to 0} L\,(\underline{M}(\xi_N) + \alpha P)^{-1}\,L^T, \tag{1.1.21}$$

where P is any positive definite matrix; see Albert (1972). Similar to $\hat{\theta}$, the estimator $\hat{\gamma}$ is b.l.u.e..

Whenever it is possible we avoid singularity of $\underline{M}(\xi_N)$. When singularity is essential for a problem, regularization similar to (1.1.20) and (1.1.21) will be used to overcome difficulties related to the pseudo-inversion technique.

Combined Estimators. Let us assume that q experiments have been performed. The corresponding l.s.e.'s are $\{\hat{\theta}_1, \ldots, \hat{\theta}_q\}$; we assume that the observations of the different experiments are uncorrelated. What is the best way to combine the information of all experiments into a single estimator?

A straightforward extension of (1.1.6) is

$$\hat{\theta} = \arg\min_{\theta} \sum_{l=1}^{q} v_l(\theta) \tag{1.1.22}$$

which is a b.l.u.e. due to the Gauss-Markov theorem. Some algebra and the use of the additivity of the information matrix lead to the combined estimator

$$\hat{\theta} = \underline{M}^{-1}(\xi_{tot}) \sum_{l=1}^{q} \underline{M}(\xi_l)\hat{\theta}_l, \tag{1.1.23}$$

where $\underline{M}(\xi_{tot}) = \sum_{l=1}^{q} \underline{M}(\xi_l)$ and ξ_{tot} is the design that comprises the support points and the corresponding weights of all individual designs, say $\xi_{tot} = \sum_{l=1}^{q} \xi_l$. The dispersion matrix of $\hat{\theta}$ is

$$\text{Var}\{\hat{\theta}\} = \underline{M}^{-1}(\xi_{tot}). \tag{1.1.24}$$

The additivity of the information matrix for combined experiments is analogous to the additivity for individual observations which are combined within one experiment.

Note that a combined estimator (1.1.23) can even be calculated if the basis functions in the models that correspond to the various experiments are different; i.e., the model structure can be different for the experiments but all of them contain the same parameters. For instance, in some experiments only the first few components of $\hat{\theta}$ can be defined, while the rest of the components may be estimated in complementary experiments. Of course, if a model does not contain a certain basis function, the information matrix for that experiment has to be extended by corresponding rows and columns of zeros in order to use (1.1.23). However, unless the combined matrix $M(\xi_{tot})$ is regular, $\hat{\theta}$ from (1.1.23) cannot be calculated.

If *a priori* information about θ is given by a vector θ_0 and the dispersion matrix D_0, then an estimator $\hat{\theta}_B$ similar to (1.1.23) is

$$\hat{\theta}_B = (\underline{M} + D_0^{-1})^{-1}(\underline{M}\hat{\theta} + D_0^{-1}\theta_0);\qquad(1.1.25)$$

the dispersion matrix is

$$\text{Var}\{\hat{\theta}_B\} = (\underline{M} + D_0^{-1})^{-1}.\qquad(1.1.26)$$

The subscript B is a hint to the obvious Bayesian interpretation of (1.1.25) and (1.1.26): If the observations y are normally distributed and the *a priori* distribution of θ is normal with expectation θ_0 and dispersion matrix D_0 then the Bayesian estimator $\hat{\theta}_B$ is the mean of the *a posteriori* distribution of θ and (1.1.26) defines the *a posteriori* dispersion matrix.

1.2 More about the Information Matrix

If the error in (1.1.1) is assumed to be random, the information matrix $\underline{M}(\xi_N)$ is the inverse of the dispersion matrix $\underline{D}(\xi_N)$. The latter matrix may be considered as a measure of the uncertainty of $\hat{\theta}$ or as a measure of the uncertainty in our knowledge of θ. "Certainty" must be substituted for "uncertainty" in the last sentence when we discuss the matrix $\underline{M}(\xi_N)$. That is the reason why $\underline{M}(\xi_N)$ is called the information matrix. Thus, both matrices $\underline{M}(\xi_N)$ and $\underline{D}(\xi_N)$ have a clear and simple meaning in the traditional setting.

The matrix $\underline{M}(\xi_N)$ (and analogously $\underline{D}(\xi_N)$) may also be used to describe the precision of the l.s.e. $\hat{\theta}$ in cases where no assumptions about the random behavior of the errors are imposed.

Local Behavior of the Sum of Squares. Writing $v(\theta)$ from (1.1.7) as

$$v(\theta) = \sum_{i=1}^{n}\sum_{j=1}^{r_i}\left(y_{ij} - \hat{\theta}^T f(x_i)\right)^2 + \sigma^2(\theta - \hat{\theta})^T\underline{M}(\xi_N)(\theta - \hat{\theta}),\qquad(1.2.1)$$

reveals the relation between the information matrix and $v(\theta)$. If we move in the parameter space from $\hat{\theta}$ a distance γ in the direction of the unit-length vector U, then

$v(\theta)$ is increased by

$$\gamma^2 U^T \underline{M}(\xi_N) U \,. \tag{1.2.2}$$

This increase is maximal if we move along the principle axis of the ellipsoid $U^T \underline{M}(\xi_N) U$ which is

$$U_{\max} = \arg \max_U U^T \underline{M}(\xi_N) U \,. \tag{1.2.3}$$

Note that

$$\lambda_{\max} = \max_U U^T \underline{M}(\xi_N) U \tag{1.2.4}$$

is the maximal eigenvalue of $\underline{M}(\xi_N)$ and $U_{\max}$ is the corresponding eigenvector. The least rate of increase will be observed along the direction

$$U_{\min} = \arg \min_U U^T \underline{M}(\xi_N) U \,. \tag{1.2.5}$$

Correspondingly,

$$\lambda_{\min} = \min_U U^T \underline{M}(\xi_N) U \tag{1.2.6}$$

is the minimal eigenvalue of $\underline{M}(\xi_N)$. Of course, if $\lambda_{\min} = 0$, we have infinitely many solutions for (1.1.6) or (1.1.8).

Another useful presentation of the information matrix can be derived from (1.2.1) to be

$$\underline{M}(\xi_N) = \frac{1}{2} \frac{\partial^2 v(\theta)}{\partial \theta \, \partial \theta^T}, \tag{1.2.7}$$

where

$$\left(\frac{\partial v}{\partial \theta} \right)^T = \frac{\partial v}{\partial \theta^T} = \left(\frac{\partial v}{\partial \theta_1}, \dots, \frac{\partial v}{\partial \theta_m} \right) \,.$$

Sensitivity of the l.s.e. with Respect to Observational Errors. Let ε_{ij} be the error of observation y_{ij}, i.e., $\varepsilon_{ij} = y_{ij} - \theta^T f(x_i)$. The derivative

$$\frac{\partial \hat{\theta}}{\partial \varepsilon_{ij}} = \sigma^{-2} \underline{M}^{-1}(\xi_N) f(x_i) \tag{1.2.8}$$

[cf. (1.1.13)] defines the relation between the l.s.e. $\hat{\theta}$ and ε_{ij} at the point $\varepsilon_{ij} = 0$. An "aggregated" sensitivity measure is

$$\sigma^2 \sum_{i=1}^n \sum_{j=1}^{r_i} \left(\frac{\partial \hat{\theta}}{\partial \varepsilon_{ij}} \right)^T \frac{\partial \hat{\theta}}{\partial \varepsilon_{ij}} = \underline{M}(\xi_N) \,. \tag{1.2.9}$$

Formulas (1.2.1), (1.2.7), and (1.2.9) give in some way a *deterministic* motivation for the use of the information matrix. Throughout almost the entire book the information matrix is a main object of discussion. The reader might have in mind the opportunity to interpret the results on the basis of this deterministic view. However, we will formulate all subsequent results in the "statistical language", i.e., we will assume that the observational errors are random.

1.3 Generalized Versions of the Linear Regression Model

Nonconstant Variance σ^2. In practice, the variance of the observational error may change for different values of the predictors, i.e.,

$$\mathrm{E}\{\varepsilon^2|x\} = \lambda^{-1}(x)\sigma^2. \tag{1.3.1}$$

If $\lambda(x)$ is known, the corresponding regression model may be reduced to the standard case by transformations

$$y\sqrt{\lambda(x)} \to y, \quad f(x)\sqrt{\lambda(x)} \to f(x). \tag{1.3.2}$$

The l.s.e. that results from (1.1.6) for the transformed data is

$$\hat{\theta} = \arg\min_{\theta} \sum_{i=1}^{n} \sum_{j=1}^{r_i} \lambda(x_i) \left(y_{ij} - \theta^T f(x_i)\right)^2; \tag{1.3.3}$$

$\lambda(x)$ is called the "weight function". The l.s.e. may be obtained either directly from (1.3.3) or by using the transformations (1.3.2) and $\hat{\theta} = \underline{M}^{-1}(\xi_N)Y$ [cf. (1.1.13)] with

$$\underline{M}(\xi_N) = \sigma^{-2} \sum_{i=1}^{n} \lambda(x_i) r_i f(x_i) f^T(x_i) \tag{1.3.4}$$

and

$$Y = \sigma^{-2} \sum_{i=1}^{n} \lambda(x_i) r_i f(x_i) \overline{y}_i. \tag{1.3.5}$$

The simplicity of transformations that are needed to handle the general case allows us to assume everywhere (with some occasional exceptions) that $\lambda(x) \equiv 1$ and to work with the shorter and simpler formulas.

Weight Function with Unknown Parameters. Often, the weight function is known but depends upon unknown parameters, i.e., $\lambda(x) = \lambda(x, \gamma)$. The general case is beyond the scope of this text and usually handled with the maximum likelihood method. An interesting and still simple case is that where the vector γ coincides with θ. A typical example is the case where the observed variable y follows the Poisson distribution so that

$$\mathrm{Var}\{y|x\} = \mathrm{E}\{y|x\} = \theta^T f(x) = \lambda^{-1}(x, \theta). \tag{1.3.6}$$

Note that $\lambda(x, \theta)$ must be positive. This implies conditional constraints for the estimation of θ.

A reasonable estimator can be obtained if guesses $\lambda_{0i} = \lambda(x_i)$, $i = 1, \ldots, n$, are available. For instance, $\lambda_{0i} = \overline{y}_i^{-1}$ might be used in the Poisson distribution case. Based on this information, an estimate of θ may be calculated as

$$\theta_1 = \arg\min_{\theta} \sum_{i=1}^{n} \lambda_{0i} p_i \left(\overline{y}_i - \theta^T f(x_i)\right)^2.$$

An obvious extension is the iterative estimation procedure based on

$$\theta_s = \arg\min_{\theta} \sum_{i=1}^{n} \lambda\,(x_i, \theta_{s-1}) p_i \left(\overline{y}_i - \theta^T f(x_i)\right)^2 , \tag{1.3.7}$$

for $s = 2, 3, \ldots$. The limit

$$\hat{\theta} = \lim_{s \to \infty} \theta_s \tag{1.3.8}$$

is called the "iterated estimator".

Two questions arise immediately: "Does the iterative procedure converge?" and "What are the statistical properties of $\hat{\theta}$?" No certain answer is possible to the first question. In practice, (1.3.8) converges in most cases. More accurately, the probability P_N that (1.3.8) converges becomes close to 1 for large N: $\lim_{N \to \infty} P_N = 1$.

If a regular matrix M exists so that

$$\lim_{N \to \infty} N^{-1} \underline{M}(\xi_N) = M , \tag{1.3.9}$$

the iterated estimator $\hat{\theta}$ is strongly consistent, i.e.,

$$a.s. \lim_{N \to \infty} \hat{\theta}_N = \theta, \tag{1.3.10}$$

where $\hat{\theta}_N$ is the iterated estimator (1.3.8) based on N observations and "$a.s. \lim$" stands for the almost sure convergence [see Rao (1973), Chpt. 2c]. Asymptotically ($N \to \infty$), the iterated estimator has the same dispersion matrix as the l.s.e. (1.3.3) with known weight function $\lambda(x)$, i.e.

$$N^{-1} \operatorname{Var}\{\hat{\theta}_N\} \simeq N^{-1} \sum_{i=1}^{n} \lambda(x_i, \theta_t) p_i f(x_i) f^T(x_i) . \tag{1.3.11}$$

It should be noted that in general

$$\hat{\theta} \neq \arg\min_{\theta} \sum_{i=1}^{n} \lambda(x_i, \theta) p_i \left(y_i - \theta^T f(x_i)\right)^2 ; \tag{1.3.12}$$

in most cases, the direct minimization leads to a biased (and inconsistent) estimator of θ. For a better understanding of the last statement the reader should consider the simple model $f^T(x) = (1, x)$ and $\lambda(x, \theta, \gamma) = (\theta_1 + \theta_2 x)^\gamma$ with $\gamma > 0$.

Multivariate Response. Let us assume that the experimenter observes more than one, say τ response variables. Then we have to extend the model (1.1.1) to

$$y = F^T(x)\theta + \varepsilon \tag{1.3.13}$$

where y and ε are τ-vectors, and m-vectors $f_i(x)$ form the rows of the $(\tau \times m)$-matrix $F(x)$:

$$F^T(x) = (f_1(x), \ldots, f_\tau(x)) .$$

Similar to the standard case, we assume that

$$E\{\varepsilon\} = 0 \text{ and } \text{Var}\{\varepsilon\} = E\{\varepsilon\varepsilon^T\} = d; \tag{1.3.14}$$

in particular, the different observations of the vector y are assumed to be uncorrelated. The l.s.e of θ is

$$\hat{\theta} = \arg\min_{\theta} v(\theta, d) \tag{1.3.15}$$

where

$$v(\theta, d) = \sum_{i=1}^{n}\sum_{j=1}^{r_i} \left(y_{ij} - F^T(x_i)\theta\right)^T d^{-1} \left(y_{ij} - F^T(x_i)\theta\right). \tag{1.3.16}$$

The minimization results in [cf. (1.1.13)]

$$\hat{\theta} = \underline{M}^{-1}(\xi_N)Y, \tag{1.3.17}$$

where

$$\underline{M}(\xi_N) = \sum_{i=1}^{n} r_i F(x_i)d^{-1}F^T(x_i), \tag{1.3.18}$$

$$Y = \sum_{i=1}^{n} r_i\, F(x_i)d^{-1}\bar{y}_i^T \tag{1.3.19}$$

$$\bar{y}_i = \frac{1}{r_i}\sum_{j=1}^{r_i} y_{ij}. \tag{1.3.20}$$

Similar to the case of univariate response the information matrix $\underline{M}(\xi_N)$ may be written as a weighted sum of information matrices of the individual observations:

$$\underline{M}(\xi_N) = \sum_{i=1}^{n} r_i\underline{M}(x_i); \tag{1.3.21}$$

however, now $1 \leq \text{rank}\,\underline{M}(x_i) \leq \tau$.

Let us note that the components of the vector y may be considered as observations of a random process taken at times $1, 2, \ldots, \tau$.

Unknown σ^2 or d. Of course, σ^2 is in most cases unknown; an unbiased estimator is

$$\hat{\sigma}^2 = (N - m)^{-1}v(\hat{\theta}). \tag{1.3.22}$$

However, most of the results in experimental design theory with one response variable do not depend upon σ^2. In contrast, in the multi-response case the knowledge of the dispersion matrix d is essential for the construction of optimal designs. Have in mind that in the case of the univariate response the l.s.e. $\hat{\theta}$ does not depend upon σ^2 [cf. (1.1.13)]; whereas in the multivariate case, the matrix d has to be known up to a multiple constant in order to calculate the l.s.e. [cf. (1.3.17)].

Unfortunately, no simple estimator exists for the case of unknown d, neither for this matrix itself nor for θ. A relatively simple estimator can be obtained in form of an iterative estimator that is calculated as follows: Starting with initial values d_0 for d, we repeat the following iteration step for $s = 1, 2, \ldots$ until a suitable stopping rule is fulfilled:

(i) given d_{s-1}, find

$$\theta_s = \arg \min_{\theta} v(\theta, d_{s-1}) ; \qquad (1.3.23)$$

(ii) calculate

$$d(\theta_s) = (N - m)^{-1} \sum_{i=1}^{n} \sum_{j=1}^{r_i} \left(y_{ij} - F^T(x_i)\theta_s \right) \left(y_{ij} - F^T(x_i)\theta_s \right)^T . \qquad (1.3.24)$$

To find θ_s, (1.3.17) may be used.

For normally distributed observations, this iterative procedure gives the maximum likelihood estimators $\hat{\theta}$ and $\hat{d}$ for θ and d, respectively:

$$\hat{\theta} = \lim_{s \to \infty} \theta_s , \quad \hat{d} = \lim_{s \to \infty} d_s . \qquad (1.3.25)$$

These estimators are strongly consistent if

$$\lim_{n \to \infty} \underline{M}(\xi_N) = \underline{M}$$

exists and is regular. It might be of interest that the maximization of the likelihood function can be transformed into the minimization

$$\min_{\theta} |d(\theta)| . \qquad (1.3.26)$$

The strong consistency of $\hat{\theta}$ and $\hat{d}$ may be established for distributions different from the normal one under rather mild assumptions about their moments.

1.4   Nonlinear Models

We generalize (1.1.1) by substituting the linear response function $\theta^T f(x)$ by a function $\eta(x, \theta)$; otherwise, the nonlinear regression model coincides with the model of the standard case (cf. Section 1.1).

Least Squares Estimation. In analogy to (1.1.8) the l.s.e. is defined as

$$\hat{\theta} = \arg \min_{\theta \in \Omega} \sum_{i=1}^{n} p_i \left(\overline{y}_i - \eta(x_i, \theta) \right)^2 , \qquad (1.4.1)$$

where Ω is a compact set in R^m. It is generally not possible to find a closed form expression for $\hat{\theta}$ like (1.1.13) in the linear case. Of course, in our computerized age we may use various optimization algorithms which allow to obtain a solution $\hat{\theta}$ of (1.4.1).

However, for assessing the quality of $\hat{\theta}$ we have to characterize the "closeness" of the estimator to the true θ_t. Unlike the estimator (1.1.13), $\hat{\theta}$ from (1.4.1) is generally biased, i.e., $\mathrm{E}\{\hat{\theta}\} \neq \theta_t$. Except for special cases, it is difficult to find an analytical expression for the dispersion matrix $\mathrm{Var}\{\hat{\theta}\}$. Even resampling techniques like boot-strapping and other Monte Carlo-type procedures are not very helpful; this is because $\mathrm{Var}\{\hat{\theta}\}$ depends on θ_t. Most of the results about $\hat{\theta}$ have asymptotical character.

Consistency of $\hat{\theta}$. Under rather mild assumptions [cf. Seber and Wild (1989), Chpt. 12] $\hat{\theta}$ is strongly consistent, i.e., (in the notation of Section 1.3)

$$a.s. \lim_{N\to\infty} \hat{\theta}_N = \theta_t \,. \tag{1.4.2}$$

The proof of consistency is based mainly on three assumptions:

(A) the response function $\eta(x,\theta)$ is continuous with respect to $\theta \in \Omega$ for all possible x;

(B) a limit $v(\theta, \theta_t)$ exists uniformly in Ω:

$$v(\theta,\theta_t) = \lim_{N\to\infty} v(\theta,\theta_t,\xi_N) = \lim_{N\to\infty} \sum_{i=1}^{n} p_{iN}[\eta(x_i,\theta) - \eta(x_i,\theta_t)]^2 \tag{1.4.3}$$

with $p_{iN} = r_i/N$;

(C) the function $v(\theta, \theta_t)$ has a unique minimum $\theta^* = \theta_t$ in Ω.

In general, the number n of different design points x_i may infinitely increase when $N \to \infty$. When $\eta(x,\theta)$ is given explicitly, verification of assumption (A) is straight-forward.

Verification of (B) is simple if, e.g., all observations are allocated at a finite number n of points $x_1, \ldots, x_n$, i.e., $\lim_{N\to\infty} p_{iN} = p_i$ for $i = 1, \ldots, n$; of course, n must be greater than the number of parameters to be estimated. In the linear case, we have

$$\begin{aligned} v(\theta,\theta_t,\xi_N) &= \sum_{i=1}^{n} \frac{r_i}{N}[\theta^T f(x_i) - \theta_t^T f(x_i)]^2 \\ &= (\theta - \theta_t)^T \sum_{i=1}^{n} \frac{r_i}{N} f(x_i)f^T(x_i)(\theta - \theta_t) \\ &= \sigma^2 N^{-1}(\theta - \theta_t)^T \underline{M}(\xi_N)(\theta - \theta_t) \,; \tag{1.4.4} \end{aligned}$$

to fulfill assumption (B) it is sufficient that the limit

$$\lim_{N\to\infty} \sigma^2 N^{-1} \underline{M}(\xi_N) = M \tag{1.4.5}$$

exists and M is regular. The consistency of $\hat{\theta}$ in the linear case is entirely determined by the behavior of ξ_N; the value of θ_t is not involved in the analysis.

The Dispersion Matrix $\mathrm{Var}\{\hat{\theta}\}$. To derive an expression for the dispersion matrix $\mathrm{Var}\{\hat{\theta}\}$ we approximate $\eta(x,\theta)$ by the power expansion in the vicinity of θ_t:

$$\eta(x,\theta) \cong \eta(x,\theta_t) + (\theta - \theta_t)f(x,\theta_t), \tag{1.4.6}$$

where
$$f(x, \theta) = \frac{\partial \eta(x, \theta)}{\partial \theta}.$$

If θ is close enough to θ_t, it may be sufficiently exact to use in (1.4.1)

$$v(\theta, \theta_t, \xi_N) = \sum_{i=1}^{n} p_i [\bar{y}_i - \eta(x_i, \theta_t) - (\theta - \theta_t) f(x_i, \theta_t)]^2. \qquad (1.4.7)$$

Comparison of (1.4.7) with (1.1.13) results in

$$\hat{\theta} - \theta_t = \underline{M}^{-1}(\xi_N, \theta_t) Y(\xi_N, \theta_t), \qquad (1.4.8)$$

where

$$\underline{M}(\xi_N, \theta_t) = \sigma^{-2} \sum_{i=1}^{n} r_i f(x_i, \theta_t) f^T(x_i, \theta_t) \qquad (1.4.9)$$

and

$$Y(\xi_N, \theta_t) = \sigma^{-2} \sum_{i=1}^{n} r_i (\bar{y}_i - \eta(x_i, \theta_t)). \qquad$$

Similarly to (1.1.15), we find

$$\text{Var}\{\hat{\theta}\} = \underline{M}^{-1}(\xi_N, \theta_t). \qquad (1.4.10)$$

Of course, (1.4.10) is valid only to the extent by which the approximation (1.4.6) is justified. Note that matrix $\underline{M}(\xi_N, \theta_t)$ depends on the unknown θ_t. At best, θ_t can be replaced by $\hat{\theta}$, so that

$$\text{Var}\{\hat{\theta}\} \cong \underline{M}^{-1}(\xi_N, \hat{\theta}). \qquad (1.4.11)$$

Iterative Estimation. Basing an iterative estimation procedure on (1.4.8) leads to the updating relation

$$\theta_{s+1} = \theta_s + \underline{M}^{-1}(\xi_N, \theta_s) \, Y(\xi_N, \theta_s). \qquad (1.4.12)$$

A modified relation that avoids extreme oscillations is

$$\theta_{s+1} = \theta_s + \alpha_s \underline{M}^{-1}(\xi_N, \theta_s) \, Y(\xi_N, \theta_s) \qquad (1.4.13)$$

where α_s is chosen so that convergence

$$\hat{\theta} = \lim_{s \to \infty} \theta_s \qquad (1.4.14)$$

is guaranteed. E.g., α_s can be found as the solution of a one-dimensional optimization problem

$$\alpha_s = \arg \min_{\alpha} \sum_{i=1}^{n} p_i (\bar{y} - \eta(x_i, \theta_s(\alpha)))^2, \qquad (1.4.15)$$

where the definition of $\theta_s(\alpha)$ is obvious from (1.4.13).

Assuming in addition to (A)-(C) the existence of the derivatives $\partial^2 \eta(x, \theta)/\partial\theta_\alpha \partial\theta_\beta$, $\alpha, \beta = 1, \ldots, m$, we can prove that

$$\lim_{N \to \infty} \sigma^{-2} N \, \text{Var}\{\hat{\theta}_N\} = M^{-1}(\theta_t), \qquad (1.4.16)$$

where
$$M(\theta_t) = \lim_{N \to \infty} \sigma^2 N^{-1} \underline{M}(\xi_N, \theta_t),$$

and that $\sigma^{-2} N \underline{M}^{-1}(\xi_N, \hat{\theta}_N)$ is a strongly consistent estimator of $M^{-1}(\theta_t)$.

Let us emphasize once more that in the nonlinear case the information matrix $\underline{M}(\xi_N, \theta_t)$ and, as a consequence, $\mathrm{Var}\{\hat{\theta}\}$ depend on the unknown θ_t.

The Maximum Likelihood Method. So far we have discussed relatively simple models like (1.1.1) or its nonlinear version, and we have avoided any assumption about the probability distribution of y beyond the existence of its mean and variance. We will continue to do that in the following chapters, leaving generalizations to the reader.

Actually, generalizations are straightforward when it is possible to show that the information which is gained in an experiment has an additive character as in (1.1.16). To see this, we assume that the i-th observation has the density function $p(y|x_i, \theta)$, $i = 1, \ldots, n$. Given sufficient regularity of $p(y|x, \theta)$ and independence of the observations, the maximum likelihood estimator

$$\hat{\theta} = \arg\min_{\theta \in \Omega} \prod_{i=1}^{n} \prod_{j=1}^{r_i} p(y_{ij}|x_i, \theta) \qquad (1.4.17)$$

is a popular choice of many practitioners. The Fisher information matrix for $\hat{\theta}$ is

$$\underline{M}(\xi_N, \theta_t) = \sum_{i=1}^{n} \sum_{j=1}^{r_i} M(x_i, \theta_t) = \sum_{i=1}^{n} r_i \underline{M}(x_i, \theta_t), \qquad (1.4.18)$$

where
$$\underline{M}(x, \theta) = \mathrm{E}\{f(y, x, \theta) f^T(y, x, \theta)\},$$

and
$$f(y, x, \theta) = \frac{\partial}{\partial \theta} \log p(y|x, \theta).$$

Thus, the Fisher information matrix has exactly the same structure as $\underline{M}(\xi_N)$ from (1.1.16). As a consequence, all results in experimental design which are given in this text and which are mainly based on (1.1.16), are valid for the maximum likelihood estimators.

Note that similar to the nonlinear case [cf. (1.4.9)] the Fisher information matrix generally depends on the unknown vector θ_t.

Chapter 2

Convex Design Theory

2.1 Optimality Criteria

From now on we will assume that only $\hat{\theta}$ and $\text{Var}\{\hat{\theta}\}$ or some functions of these quantities are used to describe the results of an experiment. This is justified by the fact that in many cases and in particular in the case of normally distributed observations, $\hat{\theta}$ and $\text{Var}(\hat{\theta})$ contain in some sense all information that is available from an experiment with respect to the linear model $y = \theta^T f(x) + \varepsilon$ [cf. Rao (1973), Chpt. 2d].

It was pointed out in Section 1.1 that for any design the l.s.e. $\hat{\theta}$ is the best one in the sense that

$$\text{Var}\{\hat{\theta}\} = \min_{\tilde{\theta}} \text{Var}\{\tilde{\theta}\}, \tag{2.1.1}$$

where minimization is with respect to all linear unbiased estimators. However, in this text we go a further step. We aim at minimizing $\text{Var}\{\hat{\theta}\}$ by choosing the most suitable design among all possible designs. This means that we have to find, due to $\text{Var}\{\hat{\theta}\} = \underline{D}(\xi_N) = \underline{M}^{-1}(\xi_N)$, the optimal design

$$\xi_N^* = \arg \min_{\xi_N} \underline{M}^{-1}(\xi_N). \tag{2.1.2}$$

Unfortunately, unlike (2.1.1), it is not possible to solve this optimization problem in general; we have to pursue much more modest objectives. In particular, we will look for

$$\xi_N^* = \arg \min_{\xi_N} \Psi \left[\underline{M}(\xi_N) \right], \tag{2.1.3}$$

where Ψ is some function that reasonably well corresponds to the needs of the experimenter. Ψ is called criterion of optimality; various functions Ψ will be discussed in the following. Obviously, an optimal design ξ_N^* depends on Ψ. From the definition of the design

$$\xi_N = \{x_i, p_i, \}_1^n$$

with $p_i = r_i/N$, it is clear that (2.1.3) is a discrete optimization problem. To find a solution may be extremely difficult both analytically and computationally. To emphasize this statement, let us rewrite (2.1.3) in the more explicit form

$$\xi_N^* = \{x_i^*, p_i^*\}_1^{n^*} = \arg \min_{x_i, p_i, n} \Psi\left[M(\{x_i, p_i\}_1^n)\right], \tag{2.1.4}$$

where $x_i \in X$ and $0 \le r_i = Np_i \le N$. The set X is called the design region. For any practical N and for even only moderately complex response models $\theta^T f(x)$, the problem (2.1.4) may be computationally too demanding even for the current thrust of computer intensive technology. Theoretical results are known only for symmetrical design regions (cubes and spheres) and for simple basis functions such as first or second order polynomials.

In the following we discuss the choice of suitable optimality criteria and approximations to (2.1.3) which allow simple and efficient construction of optimal designs. We distinguish criteria related to (a) the parameter space and to (b) the predicted response.

Popular Optimality Criteria Related to the Parameter Space. The equation

$$(\theta - \hat{\theta})^T \underline{M} (\theta - \hat{\theta}) = m \tag{2.1.5}$$

defines the so-called "ellipsoid of concentration". The "larger" the matrix $\underline{M}$ (the "smaller" $\underline{D}$) is, the "smaller" is the ellipsoid of concentration. A measure of the "size" of the ellipsoid is, for instance, its volume which is proportional to $|\underline{D}|^{1/2} = |\underline{M}|^{-1/2}$. In this text, $|A|$ stands for the determinant of matrix A. Another measure is the length of the largest principal axis of the ellipsoid which is $\lambda_{\min}^{1/2}(\underline{M})$; here, $\lambda_{\min}(\underline{M})$ stands for the minimal eigenvalue of matrix $\underline{M}$.

The figures may help to visualize the above mentioned characteristics of the information matrix. The response function is quadratic: $\theta^T f(x) = \theta_1 + \theta_2 x + \theta_3 x^2$, $\sigma^2 = 1$, and the design is $\xi_N = \{-x_1 = x_3 = 1, x_2 = 0; p_1 = p_3 = 0.25, p = 0.5\}$. In Figure 2.1, the concentration ellipses for the b.l.u.e. or weighted l.s.e. (cf. (1.1.8)) and unweighted l.s.e. are compared, when for any design all weights p_i in (1.1.8) are equal. For the same model we show in Figure 2.2 the concentration ellipses for two different designs. The first design is $\xi_{N_1} = \{-x_1 = x_3 = 1, x_2 = 0; p_1 = p_2 = p_3 = 1/3\}$; it minimizes the determinant $|\underline{D}|$. The second design is again ξ_N that is based to Figure 2.1, and it minimizes the sum of diagonal elements of $\underline{D}$, i.e., $\operatorname{tr} \underline{D}$. The last two figures give a graphical illustration of comments to (2.1.3).

Thus, we may introduce the D-criterion

$$\Psi(\underline{M}) = |\underline{M}|^{-1} = |\underline{D}| \tag{2.1.6}$$

and the E-criterion

$$\Psi(\underline{M}) = \lambda_{\min}(\underline{M}) = \lambda_{\max}(\underline{D}). \tag{2.1.7}$$

Both criteria are limiting cases of

$$\Psi_\gamma(\underline{M}) = \left(m^{-1} \operatorname{tr} \underline{M}^{-\gamma}\right)^{1/\gamma} = (m^{-1} \operatorname{tr} \underline{D}^\gamma)^{1/\gamma}; \tag{2.1.8}$$

Figure 2.1: Concentration ellipsoids for the b.l.u.e. (dashed line) and for the un-weighted l.s.e. (solid line).

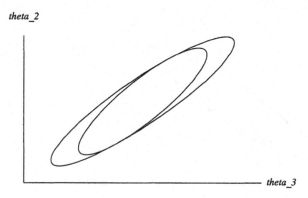

$\Psi_\gamma(\underline{M})$ results in (2.1.6) if $\gamma \to 0$, and in (2.1.7) if $\gamma \to \infty$. Another popular choice is $\gamma = 1$:

$$\Psi(\underline{M}) = m^{-1} \operatorname{tr} \underline{M}^{-1} = m^{-1} \operatorname{tr} \underline{D}. \tag{2.1.9}$$

Using Liapunov's Inequality

$$(\mathrm{E}\{|z|^\gamma\})^{1/\gamma} \le (E\{|z|^{\gamma'}\})^{1/\gamma'} \text{ for } 0 < \gamma < \gamma', \tag{2.1.10}$$

we find that

$$\begin{aligned}|\underline{D}|^{1/m} &= \lim_{\gamma \to 0}(m^{-1}\operatorname{tr}\underline{D}^\gamma)^{1/\gamma} \le (m^{-1}\operatorname{tr}\underline{D}^\gamma)^{1/\gamma} \\ &\le \lim_{\gamma \to \infty}(m^{-1}\operatorname{tr}\underline{D}^\gamma)^{1/\gamma} = \lambda_{\max}(\underline{D}).\end{aligned} \tag{2.1.11}$$

Thus, the three criteria may be ordered for any design:

$$|\underline{D}|^{1/m} \le m^{-1}\operatorname{tr}\underline{D} \le \lambda_{\max}(\underline{D}) \tag{2.1.12}$$

The D-criterion (2.1.6) is one of the most popular criteria in experimental design theory; in some textbooks the determinant $|\underline{D}(\xi_N)|$ is called the generalized variance of $\hat{\theta}$. Criterion (2.1.7) is usually called E-criterion. Criterion (2.1.9) is less easy to be used and is called A- or T-criterion [see Fedorov (1972) and Pukelsheim (1993)]. There are more criteria and these are labeled by other letters. Over the time, the

Figure 2.2: Concentration ellipsoids for the design that minimizes $|\underline{D}|$ (solid line) and for the design that minimizes tr $\underline{D}$ (dashed line).

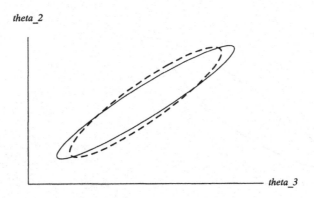

number of criteria in use grew and with it the number of letters which serve as labels; in a teasing mode, some authors speak about "alphabetic" criteria. We prefer to identify the criteria with something more informative than a letter from the Latin alphabet. However, the reader hopefully will forgive us if, for brevity, the slangish letters will appear in the text.

If we are interested in linear combinations $\zeta - L^T\theta$ of the unknown parameters where L is a (known) matrix of order $m \times s$, $s \leq m$,

$$\text{Var}\{\hat{\zeta}\} = \text{Var}\{L^T\hat{\theta}\} = L^T\underline{M}^{-1}L \qquad (2.1.13)$$

must be considered as the criterion. Although the vector ζ may be estimable even if $\underline{M}$ is singular, we always will assume that $\underline{M}$ is regular if it is not stated otherwise. The above-mentioned criteria (2.1.6), (2.1.7), and (2.1.9) may be applied to find an optimal design for estimating ζ if $\underline{M}^{-1}$ is replaced by $L^T\underline{M}^{-1}L$.

If L is an m-vector, i.e., ζ consists of only one component, a useful relationship exists between (2.1.13) and the E-criterion (2.1.7). If we look for a criterion that leads to a design which is good for any L, we would select

$$\Psi(\underline{M}) = \max_L \frac{L^T\underline{M}^{-1}L}{L^TL}. \qquad (2.1.14)$$

Taking into consideration that [cf. (1.2.3)]

$$\max_{L} \frac{L^T \underline{M}^{-1} L}{L^T L} = \lambda_{\min}(\underline{M}), \tag{2.1.15}$$

we find that (2.1.14) coincides with the E-criterion. Criteria based on (2.1.13) include the so-called linear criterion

$$\Psi(\underline{M}) = \mathrm{tr}\, A \underline{M}^{-1},$$

where $A = LL^T$ is a non-negative definite matrix that is usually called the "loss matrix".

Criteria Related to the Variance of the Predicted Response. Using (1.1.14) we can show that the variance of the predicted response at x, i.e., $\hat{\eta}(x, \theta) = \hat{\theta}^T f(x)$, is

$$\underline{d}(x) = f^T(x) \underline{M}^{-1} f(x). \tag{2.1.16}$$

Based on this we can define optimality criteria that result in designs which minimize in some sense the variance of the predicted response.

If we are interested in the response function at a particular point x_0, the choice

$$\Psi(\underline{M}) = \underline{d}(x_0) \tag{2.1.17}$$

is an obvious optimality criterion. For the case that we want to know the response function sufficiently well on some set Z, two popular criteria are

$$\Psi(\underline{M}) = \int_Z w(x) \underline{d}(x) dx \tag{2.1.18}$$

and

$$\Psi(\underline{M}) = \max_{x \in Z} w(x) \underline{d}(x); \tag{2.1.19}$$

the function $w(x)$ describes the relative importance of a response at x.

Criteria based on (2.1.16) may be considered as a special case of (2.1.13). For instance, if the integral

$$\int_Z w(x) \hat{\eta}(x, \theta) dx$$

is to be estimated, the corresponding optimality criterion is

$$\Psi(\underline{M}) = L^T \underline{M}^{-1} L \tag{2.1.20}$$

with

$$L = \int_Z w(x) f(x) dx.$$

Some Generalizations. A priori information about θ that is given by the dispersion matrix D_0 can be included in an optimality criterion by replacing the information matrix $\underline{M}$ by $\underline{M} + D_0^{-1}$ [cf. (1.1.25)].

Finally, we mention criteria that include unknown parameters. For instance, we may compare different estimates by using

$$\sum_{\alpha=1}^{m} \frac{\text{Var}\{\hat{\theta}_\alpha\}}{\theta_\alpha}.$$

The corresponding criterion is

$$\Psi(\underline{M}) = \text{tr}\, A(\theta)\underline{M}^{-1}, \qquad (2.1.21)$$

where $A(\theta)$ is a diagonal matrix with i-th diagonal element θ_i^{-1}.

One important fact makes (2.1.21) different from all other criteria that have been introduced in this section: $\Psi(\underline{M})$ is a function of the unknown parameters. Obviously, for this type of criterion it is much more difficult to construct optimal or even "good" designs than in situations where the criterion does not depend on the unknown parameters.

2.2 Some Properties of Optimality Criteria

Monotonicity. A property that is common for all optimality criteria introduced in the previous section is monotonicity: If $\underline{M} \geq \underline{M}'$, then

$$\Psi(\underline{M}) \leq \Psi(\underline{M}'), \qquad (2.2.1)$$

i.e., Ψ is a monotone function. Together with the additivity of the information matrix [cf. (1.1.16)], inequality (2.2.1) means that an additional observation cannot result in an increased value of the optimality criterion. More data can only help and never harm.

Homogeneity. When no *a priori* information is involved, then we find for all considered criteria that

$$\Psi(\underline{M}) = \Psi(N\sigma^{-2}M) = \gamma(N\sigma^{-2})\Psi(M), \qquad (2.2.2)$$

where

$$M = \sigma^2 N^{-1}\underline{M} = \sum_{i=1}^{n} p_i f(x_i) f^T(x_i) \qquad (2.2.3)$$

is the standardized information matrix as it is often called, $p_i = r_i/N$ and $\sum_{i=1}^{n} p_i = 1$, and γ is a non-increasing function. Thus, we can separate the dependence on $N\sigma^{-2}$ and M. However, matrix M still depends on N implicitly through p_i. Usually, functions of type (2.2.2) are called homogeneous functions.

Convexity. More thorough analysis shows that the considered optimality criteria are convex functions or may be transformed into convex ones; i.e.,if

$$M = (1 - \alpha)M_1 + \alpha M_2, \qquad (2.2.4)$$

for any α with $0 \leq \alpha \leq 1$, then

$$\Psi(M) \leq (1 - \alpha)\Psi(M_1) + \alpha\Psi(M_2). \tag{2.2.5}$$

Most of the criteria that have been discussed in Section 2.1 are convex.

The D-criterion $\Psi(M) = |M|^{-1}$ is not convex; but for

$$\Psi(M) = -\log|M| \tag{2.2.6}$$

convexity can be shown: We know that regular matrices M_1 and M_2 fulfill

$$|M| \geq |M_1|^{1-\alpha}|M_2|^{\alpha}$$

for M from (2.2.4) and any α with $0 \leq \alpha \leq 1$ [cf. result (A) in the Appendix A]. Convexity for (2.2.6) is an obvious corollary of this inequality.

To proof the convexity of the other criteria we make use of

$$M^{-1} \leq (1 - \alpha)M_1^{-1} + \alpha M_2^{-1}$$

where M, M_1, M_2, and α are defined as above.

2.3 Continuous Optimal Designs

A Generalized Definition of Design. We already have mentioned that it is difficult in terms both of analytical and of numerical analysis to find a solution for the discrete optimization problem (2.1.4). Each design problem requires a special approach and special efforts to find the optimal set $\{x_i^*, p_i^*\}_1^{n^*}$. Parts of these difficulties could be avoided if the discreteness of the weights p_i were ignored and the weights were assumed to be real numbers in the interval $[0, 1]$:

$$0 \leq p_i \leq 1, \quad \sum_{i=1}^{n} p_i = 1. \tag{2.3.1}$$

Corresponding designs extend the set of discrete designs so that the theory of optimization of complex functions may be applied; these designs will be called "continuous" designs.

If we have two designs ξ_1 and ξ_2 with support points $\{x_{1i}\}_1^{n_1}$ and $\{x_{2i}\}_1^{n_2}$, respectively, whose weights satisfy (2.3.1), then we may define a new design

$$\xi = (1 - \alpha)\xi_1 + \alpha\xi_2 \tag{2.3.2}$$

where $0 \leq \alpha \leq 1$. If a point x_{1i} is element only of ξ_1, it has the weight $(1 - \alpha)p_{1i}$ in ξ; if x_{2i} is element only of ξ_2, its weight in ξ is αp_{2i}. Points that are common to both designs have the weight $(1 - \alpha)p_{1i} + \alpha p_{2i}$ in ξ.

The set of points x_i in the design region X for which the design ξ has nonzero weights p_i is called the support set of ξ and denoted by

$$\operatorname{supp} \xi = \{x_i \colon p(x_i) > 0, x_i \in X\}.$$

From the definition of the weights it is obvious that any probability measure defined on the design region X can be a design; correspondingly, $\xi(dx)$ will be denoted as design measure. Appendix 2A of Rao (1973) gives a definition of probability measures and a very compact and clear description of the corresponding calculus.

Let us come back to the standardized information matrix (2.2.3). The extension of the design concept to a probability measure allows us to replace

$$M(\xi) = \sum_{i=1}^{n} p_i f(x_i) f^T(x_i) \tag{2.3.3}$$

by

$$M(\xi) = \int_X f(x) f^T(x) \xi(dx), \tag{2.3.4}$$

where the integration must be understood in the Stieltjes-Lebesgue sense. If this does not lead to misinterpretation, X will be omitted in (2.3.4). Combining (2.3.2) and (2.3.4) leads us to

$$M(\xi) = (1 - \alpha)M(\xi_1) + \alpha M(\xi_2). \tag{2.3.5}$$

Of course, (2.3.3) is a special case of (2.3.4). Those who are not familiar with measure and integration theory may interpret the integral sign as a summation in sense of (2.3.3). The use of the sum instead of the integral rarely causes problems, and there is a serious theoretical reason for it as will be seen from *Theorem 2.3.1.*

The Basic Optimization Problem. Let $\Xi(X)$ be the set of all probability measures on X. In the context of continuous designs we define an optimal design as a solution of the optimization problem

$$\xi^* = \arg \min_{\xi \in \Xi(X)} \Psi\left[M(\xi)\right]. \tag{2.3.6}$$

We omit "$\in \Xi(X)$" whenever this is not misleading. Problem (2.3.6) and its various modifications are the main content of the book. The reader will see that many experimental design problems can relatively easily be solved if they can be approximated by (2.3.6). The word "approximated" should draw your attention to the fact that a solution of (2.3.6) does generally not give an exact solution of (2.1.4). However, it is often acceptable as an approximate solution.

Alternative Optimization Problems. Similarly to the set $\Xi(X)$ of all probability measures on X we may define the set of information matrices

$$\mathfrak{M}(X) = \{M(\xi) \colon \xi \in \Xi(X)\}. \tag{2.3.7}$$

The optimization problem (2.3.6) has an analogue

$$M(\xi^*) = M^* = \arg \min_{M \in \mathfrak{M}(X)} \Psi(M). \qquad (2.3.8)$$

In theory, problem (2.3.8) is easier to solve than (2.3.6): One has to work not with probability measures but with information matrices which are elements of an $m(m+1)/2$-dimensional space. However, in some situations it is more difficult to construct $\mathfrak{M}(X)$ numerically than to solve (2.3.6). In addition, experience shows that practitioners preferably think in terms of factors (or independent variables) which must be properly selected, but seldom in terms of the mapping of $\Xi(X)$ to $\mathfrak{M}(X)$. This text is based on (2.3.6).

The Information Matrix and Properties of Optimal Designs. A rigorous mathematician would use "inf" instead of "min" in (2.3.6) and (2.3.8). This replacement allows to comprise cases where no optimal solution, i.e., an optimal point in Ξ or in $\mathfrak{M}$, exists in the traditional sense, but where structures can be found which are solutions in a generalized sense. An example for that is the notion of optimal sequences. A sequence $\{\xi_s\}$ is called optimal if the limit

$$\lim_{s \to \infty} \Psi[M(\xi_s)] = \Psi^* = \Psi[M(\xi^*)] \qquad (2.3.9)$$

exists. Such optimal sequences may be needed if the set $\mathfrak{M}(X)$ is not compact. We will avoid this situation by assuming that

(A1) X is compact, and

(A2) the basis functions $f(x)$ are continuous in X.

We will make use of a result that is usually denoted Carathéodory's Theorem [cf. Fedorov (1972), Chpt. 21, or Pukelsheim (1993), Chpt. 82]. Let S be the set of all possible $s_\mu = \int_{S_0} s\, \mu(ds)$ where μ is any probability measure defined on S_0 and $s \in S_0 \subset R^k$. Then any element of S may be represented as a convex linear combination

$$\sum_{i=1}^{n_0} \mu_i s_i, \qquad (2.3.10)$$

where $s_i \in S_0$ and $n_0 \leq k+1$. Whenever s_μ is a boundary point of S, i.e., any sphere with its center at this point that contains points both belonging and not belonging to S, then $n_0 \leq k$.

Now we can state the following theorem that summarizes our knowledge about information matrices.

Theorem 2.3.1

1. *For any design ξ the information matrix $M(\xi)$ is symmetric and non-negative definite.*

2. The set $\mathfrak{M}(X)$ is compact and convex.

3. For any matrix M from $\mathfrak{M}(X)$ exists a design ξ that contains not more than

$$n_0 = \frac{m(m+1)}{2} + 1$$

points with non-zero weight and $M(\xi) = M$. If M is a boundary point of $\mathfrak{M}(X)$, $n_0 = m(m+1)/2$.

Proof: 1. Symmetry of the information matrix $M(\xi)$, i.e., $M(\xi) = M^T(\xi)$, follows from definition (2.3.4). For any vector a, we get

$$a^T M(\xi) a = \int a^T f(x) f^T(x) a \, \xi(dx) = \int \left(a^T f(x) \right)^2 \xi(dx) \geq 0 \, ;$$

this proves that $M(\xi)$ is non-negative definite.

2. Compactness of $\mathfrak{M}(X)$ is a direct consequence of assumptions (A1) and (A2). Convexity is stated in (2.3.5).

3. $M(\xi)$ is defined to be

$$M(\xi) = \int M(x) \xi(dx),$$

where $M(x) = f(x) f^T(x)$. In other words, the set $\mathfrak{M}(X)$ is a convex hull of the set $\{M(x) : x \in X\}$. Due to the symmetry of $M(x)$, the actual dimension of this set is $k = m(m+1)/2$. Consequently, Carathéodory's Theorem states that any element from $\mathfrak{M}(X)$ may be represented as a convex linear combination of not more than n_0 elements $M(x_i)$ [see the comments to (2.3.10)]:

$$M(\xi) = \sum_{i=1}^{n_0} p_i M(x_i), \quad \sum_{i=1}^{n_0} p_i = 1 \, ,$$

where $n_0 \leq k+1$ in general, and $n_0 \leq k$ for boundary points. The choice of $\{x_i, p_i\}_1^n$ for ξ completes the proof.

Part 3 of *Theorem 2.3.1* is of great practical relevance. It allows us to restrict the search of optimal design to designs with just n_0 support points. Therefore, those who are reluctant to exercise with Stieltjes-Lebesgue integrals may think in terms of designs with a finite number of support points.

Whenever it is not explicitly stated otherwise, we shall assume that assumptions (A1) and (A2) are fulfilled, so that the validity of *Theorem 2.3.1* is assured.

A Necessary and Sufficient Condition for Optimality. Now we state four assumptions for the optimality criterion $\Psi(M)$.

(B1) $\Psi(M)$ is a convex function, i.e., for any α with $0 \leq \alpha \leq 1$,

$$\Psi\left[(1-\alpha) M_1 + \alpha M_2 \right] \leq (1-\alpha)\Psi(M_1) + \alpha \Psi(M_2) \, ; \qquad (2.3.11)$$

(B2) $\Psi(M)$ is a monotone function, i.e., for any matrix $\Delta \geq 0$,

$$\Psi(M) \leq \Psi(M + \Delta);$$

(B3) a real number q can be found such that the set

$$\{\xi \colon \Psi[M(\xi)] \leq q < \infty\} = \Xi(q)$$

is not empty;

(B4) for any $\xi \in \Xi(q)$ and $\overline{\xi} \in \Xi$,

$$\Psi\left[(1 - \alpha)M(\xi) + \alpha M(\overline{\xi})\right] \qquad (2.3.12)$$

$$= \Psi[M(\xi)] + \alpha \int \psi(x, \xi)\overline{\xi}(dx) + o(\alpha|\xi, \overline{\xi}),$$

where the function o is so that $\lim_{\alpha \to 0} \alpha^{-1} o(\alpha|\xi, \overline{\xi}) = 0$.

Assumptions (B1) and (B2) do not contain anything that is new or restrictive: They refer to convexity and monotonicity as discussed in Section 2.2, where we found that all popular criteria satisfy this assumption. In (B3) we modestly assume the existence of designs that have a finite value of the optimality criterion. The most restrictive assumption is probably (B4). It states the existence of the directional derivative

$$\left.\frac{\partial \Psi\left[(1 - \alpha)M(\xi) + \alpha M(\overline{\xi})\right]}{\partial \alpha}\right|_{\alpha=0} = \lim_{\alpha \to 0} \frac{(1 - \alpha)\Psi[M(\zeta)] + \alpha\Psi[M(\overline{\xi})]}{\alpha}$$

for any $\alpha \geq 0$. Moreover, this derivative must admit a rather specific form

$$\int \psi(x, \xi)\overline{\xi}(dx).$$

The meaning of $\psi(x, \xi)$ will be discussed and interpreted below.

Most of criteria discussed in Section 2.2 satisfy (B4). Unfortunately, the E-criterion (2.1.7) and the criterion (2.1.19) that is related to the variance of the predicted response do in general not belong to this group. Assumption (B4) may be violated in cases where optimal designs happen to be singular, i.e., rank $M(\xi^*) < m$. Examples for that might occur in the context of the linear criterion (2.1.13) if rank$(L) < m$; see Ermakov (1983) and Pukelsheim (1993) for details.

We will use short forms $\Psi(\xi)$, Ψ^*, $\min_x$, $\min_\xi$, $\int$, etc. instead of $\Psi[M(\xi)]$, $\Psi(\xi^*)$, $\min_{x \in X}$, $\min_{\xi \in \Xi}$, $\int_X$, respectively, if this does not lead to ambiguities.

Theorem 2.3.2 *Suppose that assumptions (A1)-(A2) and (B1)-(B4) hold; then the following statements are true:*

1. *There exists an optimal design ξ^* that contains not more than $m(m + 1)/2$ support points.*

2. *The set of optimal designs is convex.*

3. *A necessary and sufficient condition for a design ξ^* to be optimal is the inequality*

$$\min_x \psi(x, \xi^*) \leq 0. \qquad (2.3.13)$$

4. *The function $\psi(x, \xi^*)$ has the value zero almost everywhere in supp ξ^*.*

Proof: 1. The existence of an optimal design follows from the compactness of $\mathfrak{M}$ (cf. *Theorem 2.3.1*) and (B3). Because of the monotonicity of Ψ, $M(\xi^*)$ must be a boundary point of $\mathfrak{M}$. From part 3 of *Theorem 2.3.1* follows the existence of an optimal design with not more than $m(m+1)/2$ support points.

2. Let ξ_1^* and ξ_2^* be optimal, i.e., $\Psi[M(\xi_1^*)] = \Psi[M(\xi_2^*)] = \min_\xi \Psi[M(\xi)]$ and $\xi^* = (1-\alpha)\xi_1^* + \alpha\xi_2^*$. Then by (B1),

$$\Psi[M(\xi^*)] \leq (1-\alpha)\Psi[M(\xi_1^*)] + \alpha\Psi[M(\xi_2^*)] = \min_\xi \Psi[M(\xi)],$$

i.e., ξ^* is optimal.

3. If $\Psi[M(\xi)]$ is a convex function of ξ, then the non-negativity of the directional derivative at ξ^* is a necessary and sufficient condition for the optimality of ξ^*. From this fact and (B4), we may conclude that it is a necessary and sufficient condition for the optimality of ξ^* that ξ^* fulfills the inequality

$$\min_\xi \int \psi(x, \xi^*)\xi(dx) \geq 0, \qquad (2.3.14)$$

and consequently

$$\min_x \psi(x, \xi^*) \geq 0.$$

4. Let us assume that a subset $X' \subset$ supp ξ^* exists such that

$$\int_{X'} \psi(x, \xi^*)\xi^*(dx) \leq \gamma$$

for some $\gamma < 0$ and that $\psi(x, \xi^*) = 0$ everywhere in supp $\xi^* \setminus X'$. Then

$$\int_X \psi(x, \xi^*)\xi^*(dx) \leq \gamma. \qquad (2.3.15)$$

However, (2.3.15) contradicts

$$\int_X \psi(x, \xi^*)\xi^*(dx) = 0,$$

which is an obvious consequence of (B4) for $\overline{\xi} = \xi^*$. This contradiction proves the concluding part of *Theorem 2.3.2*.

It is useful to note that the following inequality holds for any optimality criterion which satisfies the conditions of *Theorem 2.3.2* and for any design ξ:

$$\min_x \psi(x, \xi) \leq \Psi[M(\xi^*)] - \Psi[M(\xi)]; \qquad (2.3.16)$$

ξ^* is an optimal design. To verify this inequality, we consider the design $\bar{\xi} = (1 - \alpha)\xi + \alpha\xi^*$ and make use of the convexity of Ψ, i.e., of assumption (B1):

$$\Psi[M(\bar{\xi})] \leq (1 - \alpha)\Psi[M(\xi)] + \alpha\Psi[M(\xi^*)].$$

From that follows that

$$\frac{\Psi[M(\bar{\xi})] - \Psi[M(\xi)]}{\alpha} \leq \Psi[M(\xi^*)] - \Psi[M(\xi)].$$

Consequently, when $\alpha \to 0$ [see (B4)],

$$\int \psi(x, \xi)\xi(dx) \leq \Psi[M(\xi^*)] - \Psi[M(\xi)].$$

Observing that

$$\int \psi(x, \xi)\xi(dx) \geq \min_x \psi(x, \xi)$$

leads us to (2.3.16).

2.4 The Sensitivity Function and Equivalence Theorems

The Sensitivity Function. The function $\psi(x, \xi)$ plays a major role in convex design theory. For instance, from part 4 of *Theorem 2.3.2* follows that $\psi(x, \xi^*)$ completely determines the location of the support points of the optimal design ξ^*. This fact together with (B4) makes clear that $\psi(x, \xi^*)$ indicates points where taking an observation provides most information with respect to the chosen optimality criterion. We also can see that moving some small measure from the support set of ξ to some point x decreases $\Psi(\xi)$ by $-\alpha\psi(x, \xi)$. In other words, $\psi(x, \xi)$ indicates points at which an observation contributes most to approaching the optimal design.

This is probably the reason why the function $\phi(x, \xi) = -\psi(x, \xi) + C$ is called the "sensitivity function". The constant C is equal to the number of estimated parameters for the D- and related criteria and equal to $\Psi(\xi)$ for all other criteria considered so far. *Table 2.1* shows the sensitivity function for the most popular optimality criteria. For the sake of simplicity we assume that rank $M(\xi^*) = m$, i.e., ξ^* is regular. There exists a number of design problems where this is not true [cf. Pukelsheim (1993)]. However, in many cases simple regularization procedures allow to get practical solutions that are close to singular optimal designs in the sense of the corresponding criteria. To make the expressions in the second column of *Table 2.1* simpler, we dropped in some cases factors that do not depend upon x.

Equivalence Theorems. It has a long tradition in experimental design theory to make use of so-called "equivalence theorems" that are obvious modifications of *Theorem 2.3.2* for various optimality criteria. The most celebrated one is the Kiefer-Wolfowitz equivalence theorem [see, e.g., Karlin and Studden (1966) or Fedorov (1972)]:

Table 2.1: Sensitivity function $\phi(x, \xi)$ for various optimality criteria $\Psi(\xi)$.

$\Psi(\xi)$	$\phi(x,\xi)$	C
$\log \lvert D\rvert,\ D = M^{-1}$	$d(x) = f^T(x)Df(x)$	m
$\log \lvert D_\ell\rvert$ *	$d(x) - d_k(x)$ $d_k(x) = f_k^T(x)D_k f(x),$ $f^T(x) = (f_\ell^T(x), f_k^T(x))$	k
$\operatorname{tr} AD,\ A \geq 0$	$f^T(x)DADf(x)$	$\operatorname{tr} AD$
$d(x_0)$	$d^2(x, x_0)$ $d(x, x_0) = f^T(x)Df(x_0)$	$d(x_0)$
$\int_Z d(x)dx$	$\int_Z d^2(x, z)dz$	$\int_Z d(x)dx$
$\operatorname{tr} D^\gamma$	$f^T(x)D^{\gamma+1}f(x)$	$\operatorname{tr} D^\gamma$
$\lambda_{\min} = \lambda_{\min}(M)$ $= \lambda_{\max}^{-1}(D)$	$\sum_{i=1}^\alpha \pi_i(f^T(x)P_i)^2$ $\lambda_{\min}P_i = MP_i,$ α is a multiple of $\lambda_{\min},$ $\sum_{i=1}^\alpha \pi_i = 1,\ 0 \leq \pi_i \leq 1$	$\lambda_{\min}$

* D_ℓ is a submatrix of D corresponding to ℓ parameters, $k = m - \ell$.

Theorem 2.4.1 *The following design problems are equivalent: Find ξ so that*

1. $\min_\xi \lvert D(\xi)\rvert,$

2. $\min_\xi \max_x d(x, \xi),$

3. $\max_x d(x, \xi) = m.$

To see the relation of this theorem with *Theorem 2.3.2* we choose $\Psi = \log \lvert D\rvert$ and $\psi(x, \xi) = m - d(x, \xi)$. Recalling that

$$\begin{aligned}
\operatorname{Var}\{f^T(x)\hat\theta\} &= f^T(x)\underline{M}^{-1}(\xi)f(x) \\
&= \sigma^2 N^{-1}f^T(x)D(\xi)f(x) = \sigma^2 N^{-1}d(x, \xi),
\end{aligned}$$

we see that observations in a D-optimal design must be taken at points where the variance of the predicted response function is largest. This fact is in a good agreement with intuition.

Figure 2.3 illustrates *Theorem 2.4.1*. The ure shows the variance of the response function $d(x, t, \xi)$ of a polynomial regression of order 4 on $X = [-1, 1]$ for two designs: (i) a uniform design where 101 observations are equally spaced over X, and (ii) the D-optimal design which allocates the observations with equal weights at $x = \pm 1$, ± 0.65, and 0. For $\lvert D(\xi)\rvert$, the values 209.0 10^3 and 23.2 10^3 correspond to the two designs.

Figure 2.3: Polynomial regression of order 4: Sensitivity function $d(x, \xi)$ for a uniform design and for the D-optimal design (bold line).

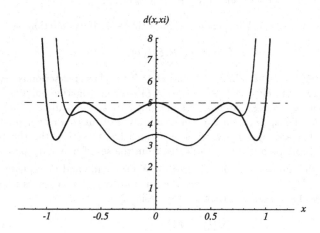

is of practical interest in its own. We are looking for a design which minimizes the maximum variance of the estimated response for all points in X. Two facts should be emphasized. First, the maximum is taken over the whole design region X [cf. (2.1.19)]. Second, it is assumed that all observations have the same variance. Otherwise, (2.4.1) must be replaced by

$$\xi^* = \arg \min_\xi \max_x \lambda(x) d(x, \xi) . \qquad (2.4.2)$$

where $\lambda(x)$ is defined in (1.3.1).

Theorems similar to *Theorem 2.4.1* may be formulated for other criteria.

Theorem 2.4.2 *The following design problems are equivalent: Find ξ so that*

1. $\min_\xi \Psi(\xi)$,

2. $\min_\xi \max_x \phi(x, \xi)$,

3. $\max_x \phi(x, \xi) = \Psi(\xi)$.

2.5   Some Examples

Simple Linear Regression. We are interested in the model $y = \eta(x, \theta) + \varepsilon$ with

$$\eta(x, \theta) = \theta_1 + \theta_2 x_1 + \ldots + \theta_m x_{m-1} \,,$$

i.e., $f^T(x) = (1, x_1, \ldots, x_{m-1})$. The general form of the sensitivity function is

$$\phi(x, \xi) = f^T(x) B(\xi) f(x)$$

where the matrix $B(\xi)$ stands for D, DAD, and $D^{\gamma+1}$ if the chosen criterion is $\log |D|$, $\operatorname{tr} AD$, and $\operatorname{tr} D^\gamma$, respectively (cf. *Table 2.1*); $\phi(x, \xi)$ is a non-negative polynomial of second order. Obviously, $\phi(x, \xi)$ can reach its maxima only at the boundary points of the design region X. Thus, *Theorem 2.3.2* immediately leads us to a rather practical recommendation. For any optimality criterion (which has to satisfy the theorem's conditions), the support points must be selected from the set of boundary points.

Let us discuss this for two popular choices for X: the cube and the sphere. For the cube, i.e., when $|x_{i-1}| \leq 1$, $i = 2, \ldots, m$, the function $\phi(x, \xi)$ reaches its maxima only at the vertices. Let us assume that for any design ξ

$$\Psi(\xi) = \Psi(\underline{\xi}) \tag{2.5.1}$$

where ξ and $\underline{\xi}$ differ only by the signs of some (or all) coordinates of their support points. Not all criteria satisfy this assumption; an example for this fact is $\operatorname{tr} AD$ for some matrices A. But for "good" ones convexity [see assumption (B1)] leads to the inequality

$$\Psi(\xi_c) \leq \Psi(\xi) = \Psi(\underline{\xi}) \,,$$

where $\xi_c = \frac{1}{2}\xi + \frac{1}{2}\underline{\xi}$. From this fact follows that a design that contains all vertices as support points with equal weight is optimal. Thus, we have obtained an optimal design that has a very simple and intuitively obvious structure; it is optimal in the sense of any optimality criterion for which *Theorems 2.3.2* and *2.4.2* are true. Unfortunately, this design contains an exponentially increasing number $n = 2^m$ of support points. A long time concern in experimental design theory is the question how this number can be reduced. The assumptions that the cube is centered at 0 and that $|x_{i-1}| \leq 1$ for all i is not essential if the optimality criterion is invariant with respect to non-singular linear transformations.

In the case of a sphere an optimal design may be constructed similar to the cube if we assume invariance with respect to rotation instead of (2.5.1). We immediately find that the sensitivity function must be constant on the boundary. This fact indicates that a design with uniform measure on the boundary sphere is optimal. Another optimal design has support points where the coordinate axes pierce the sphere and all have the same weight $1/2(m-1)$. Both designs are theoretically simple but they are not very attractive for the practitioner. Evidently, the design that consists of support points with equal weights m^{-1} which coincide with the vertices of any regular simplex that is inscribed in the boundary sphere is also optimal.

The case where X is a sphere provides a good and simple example of a singular optimal design. If $\Psi(\xi) = d(x_0, \xi)$, the design problem can be reduced to a one-dimensional problem. All candidate points lie on the line that connects the center of the sphere and the point x_0. The optimal design has two support points where this "connecting" line pierces the sphere. If x_0 lies outside of the sphere, i.e., $r^2 = x_0^T x_0 > 1$, the weight of the support point that is closest to x_0 is $1/2(1+r)$ [cf. Fedorov (1972), Chpt. 2].

Polynomial Regression on the Interval $[-1, 1]$. The polynomial regression with $f^T(x) = (1, x, \ldots, x^{m-1})$ and $X = [-1, 1]$ is one of the most frequently used test grounds for various exercises in experimental design theory; see Karlin and Studden (1966), Fedorov (1972), or Pukelsheim (1993).

Let us consider the prediction of the response at point x_0. To avoid singular designs we assume that $x_0 > 1$. From *Theorem 2.4.2* follows that for an optimal design (see *Table 2.1*)

$$\max_x |d(x, x_0, \xi^*)| = \sqrt{d(x_0, \xi^*)} \qquad (2.5.2)$$

where the sensitivity function $\phi(x, \xi) = f^T(x)D(\xi)f(x) = d^2(x, x_0, \xi)$ and $d(x_0, \xi)$ is the variance of the predicted response, respectively, given a design ξ. The function $d(x, x_0, \xi)$ is a polynomial of degree $m - 1$ and problem (2.5.2) looks very similar to the search of the polynomial that deviates least from zero. We may guess that the support points coincide with a Tchebysheff support, i.e.,

$$x_i^* = \cos\left(\frac{m-i}{m-1}\pi\right) \qquad (2.5.3)$$

for $i = 1, \ldots, m$, and $\phi(x, \xi^*) = C(x_0)T_{m-1}(x)$, where $C(x_0)$ depends only on the position of x_0 and $T_{m-1}(x) = \cos[(m-1)\arccos x]$ is the Tchebysheff polynomial [see Fedorov (1972), Karlin and Studden (1966), Szegö (1959)]. If a support set contains exactly $n = m$ points we can find that $d(x, \xi) = \sum p_i^{-1} L_i^2(x)$, and

$$d^2(x, x_0, \xi) = \sum p_i^{-1} L_i(x)L_i(x_0), \qquad (2.5.4)$$

where

$$L_i(x) = \frac{\prod_{j\neq i}(x - x_j)}{\prod_{j\neq i}(x_i - x_j)}$$

for all i. Combining (2.5.2) with (2.5.4) gives

$$p_i^* = \frac{|L_i^*(x_0)|}{\sum_{j=1}^m |L_j^*(x_0)|} \qquad (2.5.5)$$

for all i where the asterisk indicates that $\{x_i^*\}_1^m$ coincides with the Tchebysheff basis.

If the weights are selected according to (2.5.5), the property (2.5.2) is satisfied. Moreover, the sensitivity function $\phi(x, \xi^*)$ has the same value for all x_i^*. Therefore, $\phi(x, \xi^*)$ must be proportional to $T_{m-1}(x)$. Thus, (2.5.3) and (2.5.5) define the optimal design for the prediction of the response at x_0.

Note that the D-optimal design for this model has a support set that coincides with the roots of the polynomial $(1-x^2)dP_{m-1}(x)/dx$, where $P_{m-1}(x)$ is the $(m-1)$-st Legendre polynomial [see Fedorov (1972)].

Trigonometric Regression. The following example illustrates an approach that often is successfully applied. We start with some guess for the optimal design and then use *Theorem 2.3.2* or *2.4.2* in order to verify (or falsify) that this guess is correct. The model is defined by

$$f^T(x) = (1, \sin x, \cos x, \ldots, \sin kx, \cos kx)$$

where $0 \le x < 2\pi$, and contains $2k + 1$ parameters. Intuition implies that for an optimality criterion that is invariant with respect to rotation, the corresponding design must look similar in coordinate systems that differ only in the origin of x. Let us choose the uniform design (measure) on $[0, 2\pi)$ as our first candidate for the optimal design: $\xi^*(dx) = 1/2\pi$. We will verify that this design is D-optimal; we could similarly proceed for other criteria.

For any integers α, $\beta \le k$ we know that

$$\int_0^{2\pi} \sin \alpha x \, dx = \int_0^{2\pi} \cos \alpha x \, dx = \int_0^{2\pi} \cos \alpha x \sin \beta x \, dx = 0,$$

if $\alpha \ne \beta$, and

$$\frac{1}{2\pi} \int_0^{2\pi} \sin^2 \alpha x \, dx = \frac{1}{2\pi} \int_0^{2\pi} \cos^2 \alpha x \, dx = \frac{1}{2}. \tag{2.5.6}$$

From (2.5.6) follows that

$$D(\xi^*) = \begin{pmatrix} 1 & 0 & \ldots & 0 \\ 0 & 2 & \ldots & 0 \\ \vdots & \vdots & \ddots & \vdots \\ 0 & 0 & \ldots & 2 \end{pmatrix} \tag{2.5.7}$$

and

$$\begin{aligned} d(x, \xi^*) &= f^T(x)D(\xi^*)f(x) = 1 + 2(\sin^2 x + \cos^2 x) + \ldots \\ &\quad + 2(\sin^2 kx + \cos^2 kx) \equiv 2k + 1 = m \, ; \end{aligned} \tag{2.5.8}$$

from *Theorem 2.4.1* is clear that the design ξ^* is D-optimal. Repeating the derivation of (2.5.8) for matrix $D^{\gamma+1}$ gives (see *Table 2.1*)

$$\begin{aligned} \phi(x, \xi^*) &= f^T(x)D^{\gamma+1}f(x) \\ &= 1 + 2^{\gamma+1}(\sin^2 x + \cos^2 x) + \ldots + 2^{\gamma+1}(sin^2 kx + \cos^2 kx) \\ &= 2^{\gamma+1}k + 1 \end{aligned}$$

and

$$\operatorname{tr} D^\gamma = 2^\gamma 2k + 1 = 2^{\gamma+1}k + 1 \, ;$$

the design ξ^* is optimal in the sense of (2.1.8) for any real γ.

Mathematically the uniform design is a perfect object to be analyzed. However, for a practitioner a design with infinitely many support points is not the most desirable thing. Fortunately, any design ξ_n^* with support points

$$x_i = 2\pi \frac{i-1}{n} + \varphi$$

for $i = 1, \ldots, n$, $n \geq m$, and equal weights has a matrix $D(\xi_n^*)$ which coincides with (2.5.7), and therefore ξ_n^* is optimal with respect to the same optimality criteria as ξ^*. The constant φ has the effect of a phase and may be any real number. This means that the design points $\{x_i\}_1^n$ can simultaneously be rotated. Actually, it is more convenient to consider this circle as the design region X than the interval $[0, 2\pi)$.

2.6 Complements

Regularization. We have already mentioned that optimal designs may be singular, i.e., they have a singular information matrix, the rank $M(\xi) < m$. In this situation we replace the inverse by the generalized inverse of M if we have to estimate the parameters; cf. what is said in the context of (1.1.19). The problem is essentially more difficult in our situation. The limit transition technique [cf. (1.1.20) and (1.1.21)] does not work properly. In general, the singular case needs more sophisticated mathematics and is beyond the scope of this book. The following modification of the original optimization problem can be helpful.

Singularity of M causes problems when we have to differentiate $\Psi(M)$. The assumption (B4) states the existence of the directional derivative of $\Psi(M)$ (see Section 2.3). To cope with (B4) in cases of singular optimal designs let us introduce

$$\Psi_\gamma(\xi) = \Psi\left[(1-\gamma)M(\xi) + \gamma M(\xi_0)\right], \qquad (2.6.1)$$

where $0 < \gamma < 1$ and rank $M(\xi_0) = m$. If assumption (B4) is fulfilled for $\Psi(M)$ when M is regular, then it is valid for $\Psi_\gamma(\xi)$ whatever the design ξ is. If

$$\xi_\gamma^* = \arg\min_\xi \Psi_\gamma(\xi), \qquad (2.6.2)$$

then from the convexity of $\Psi_\gamma(\xi)$ immediately follows that

$$\Psi(\xi_\gamma^*) - \Psi(\xi^*) \leq \gamma\left[\Psi(\xi_0) - \Psi(\xi^*)\right]. \qquad (2.6.3)$$

The proper choice of γ and ξ_0 may assure "practical" optimality of ξ_γ^*.

The optimization problem (2.6.2) is a special case of (2.3.6) where all necessary assumptions including (B4) hold. Therefore, the *Theorems 2.3.2* and *2.4.2* work. E.g., for the linear criterion $\Psi(M) = \operatorname{tr} AM^{-1}$ (in the singular case $\Psi(M) = \operatorname{tr} AM^-$) the sensitivity function is

$$\phi(x, \bar{\xi}) = f^T(x)M^{-1}(\bar{\xi})AM^{-1}(\bar{\xi})f(x)$$

and
$$C = \text{tr } AM^{-1}(\overline{\xi})M(\xi)M^{-1}(\overline{\xi})$$
where $\overline{\xi} = (1-\gamma)\xi + \gamma\xi_0$.

Bayesian Optimal Design. If *a priori* information is available in the form of an *a priori* distribution with dispersion matrix $\underline{D}_0$, then [see (1.1.26)] $\underline{M}_{tot}(\xi) = \underline{M}(\xi) + \underline{D}_0^{-1}$ and in the normalized version

$$M_{tot}(\xi) = M(\xi) + D_0^{-1} = M(\xi) + M_0, \tag{2.6.4}$$

where $D_0 = \sigma^{-2}N\underline{D}_0 = M_0^{-1}$. Let us consider the optimal design

$$\xi_B^* = \arg\min_\xi \Psi[M(\xi) + M_0] \tag{2.6.5}$$

that takes the *a priori* information into account.

Compared to (2.3.6) nothing very special can be seen in (2.6.5). For instance, the D-optimal design is defined to be

$$\xi_B^* = \arg\max_\xi |M(\xi) + M_0|. \tag{2.6.6}$$

We find that

$$\begin{aligned}
\psi(x,\xi) &= \text{tr } M_{tot}^{-1}(\xi)M(\xi) - f^T(x)M_{tot}^{-1}(\xi)f(x) \\
&= \text{tr } M_{tot}^{-1}(\xi)M(\xi) - d_{tot}(x,\xi),
\end{aligned} \tag{2.6.7}$$

where $d_{tot}(x,\xi)$ is the normalized variance of the predicted response function. For the sensitivity function and for C (cf. *Table 2.1*) we get

$$\phi(x,\xi) = d_{tot}(x,\xi), \quad C = \text{tr } M_{tot}^{-1}(\xi)M(\xi) < m. \tag{2.6.8}$$

In spite of the seemingly entire analogy there is one fact which makes (2.6.6) and (2.3.6) very different for application: In the Bayesian case the optimal design ξ_B^* in general depends on the variance σ^2 and the number of observations N in the newly designed experiment.

The Optimality Criterion Depends on Uncontrolled Variables. Let us consider the design problem where the optimality criterion $\Psi(M, u)$ depends on variables or parameters $u \in U$ that cannot be controlled by the experimenter. An example is the criterion (2.1.21) that depends on values θ_α, $\alpha = 1, \ldots, m$ (cf. Section 2.1). We assume that the optimality criterion $\Psi(M, u)$ satisfies the assumptions (A1)-(A2) and (B1)-(B4) for every $u \in U$. In such a situation it is generally impossible to find the design that is best for all possible u. Therefore, we will discuss concepts that may be called "optimal design in average" and "minimax design". A design that is optimal in average is defined as

$$\xi_A^* = \arg\min_\xi \int_U \Psi[M(\xi), u]A(du) = \arg\min_\xi \Psi_A[M(\xi)] \tag{2.6.9}$$

where $\int_U A(du) = 1$; $A(du)$ may be interpreted as a measure of our trust in a particular value of u. In the Bayesian approach, such a measure is considered as an *a priori* distribution of u.

The minimax criterion defines the optimal design as

$$\xi_M^* = \arg \min_\xi \max_u \Psi[M(\xi), u] = \arg \min_\xi \Psi_M[M(\xi)]. \qquad (2.6.10)$$

Optimality in Average. We assume that U is compact and that $A(du) = a(u)du$. The design ξ_A^* often is called Bayesian. We prefer to denote ξ_A^* an optimal design in average and not a Bayesian design in order to avoid a mix-up with ξ_B^* [cf. (2.6.5)] for which the label Bayesian should be reserved.

The optimization problem (2.6.9) is relatively simple. Assumption (B4) defines the function $\psi(x, u, \xi)$ for criterion $\Psi(M, u)$. As integration is a linear operation we can reformulate the *Theorems 2.3.2* and *2.4.2* for $\Psi_A(M)$ by introducing

$$\psi(x, \xi) = \int_U \psi(x, u, \xi) A(du) \qquad (2.6.11)$$

and

$$\phi(x, \xi) = \int_U \phi(x, u, \xi) A(du). \qquad (2.6.12)$$

The Minimax Design. To find a solution for the minimax problem (2.6.10) we restate assumption (B4):

(B4) For any $\xi \in \Xi(q)$ and $\bar{\xi} \in \Xi$

$$\Psi\left[(1 - \alpha)M(\xi) + \alpha M(\bar{\xi}), u\right] = \Psi[M(\xi), u]$$
$$+ \alpha \int \psi(x, \xi, u)\bar{\xi}(dx) + o(\alpha|\xi, \bar{\xi}), \qquad (2.6.13)$$

where $\lim_{\alpha \to 0} \alpha^{-1} o(\alpha|\xi, \bar{\xi}) = 0$ uniformly in U.

We make use the of relation

$$\frac{\partial}{\partial \alpha} \max_{u \in U} g(\alpha, u) = \max_{u \in U^*} \frac{\partial g(\alpha, u)}{\partial \alpha}, \qquad (2.6.14)$$

which is true if U is compact and if the corresponding derivative exists everywhere in U [cf. Pshenichnyi (1971), Chpt. 3]. The set U^* is defined as

$$U^* = \{u : u^* = \arg \max_{u \in U} g(\alpha, u)\}.$$

We get

$$\frac{\partial}{\partial \alpha} \max_{u \in U} \Psi\left[(1 - \alpha)M(\xi) + \alpha M(\bar{\xi}), u\right]_{\alpha=0} =$$
$$= \max_{u \in U(\xi)} \frac{\partial}{\partial \alpha} \Psi\left[(1 - \alpha)M(\xi) + \alpha M(\bar{\xi}), u\right]_{\alpha=0}$$
$$= \max_{u \in U(\xi)} \int \psi(x, u, \xi)\bar{\xi}(dx), \qquad (2.6.15)$$

where
$$U(\xi) = \{u : u(\xi) = \arg\max_{u \in U} \Psi[M(\xi), u]\}.$$

Nonlinearity of the "maximization" procedure leads to the necessary and sufficient condition of optimality which is more complicated than that in *Theorems 2.3.2* and *2.4.2*:

Theorem 2.6.1 *A necessary and sufficient condition for ξ_M^* to be optimal is the existence of a measure ζ^* such that*

$$\min_x \psi(x, \xi_M^*, \zeta^*) \geq 0, \tag{2.6.16}$$

where

$$\psi(x, \xi, \zeta) = \int_{U(\xi)} \psi(x, u, \xi)\, \zeta(du). \tag{2.6.17}$$

Proof: The proof is analogous to the corresponding part of the proof of *Theorem 2.3.2*. The difference is that we use (2.6.15) and the fact that

$$\min_{\bar\xi} \max_{\zeta} \int_X \int_{U(\xi)} \psi(x, u, \xi)\bar\xi(dx)\, \zeta(du)$$

$$= \min_{\bar\xi} \max_{u \in U(\xi)} \int_X \psi(x, u, \xi)\, \bar\xi(dx)$$

$$= \max_{\zeta} \min_x \int_{U(\xi)} \psi(x, u, \xi)\, \zeta(du)$$

$$= \max_{\zeta} \min_{\bar\xi} \int_X \int_{U(\xi)} \psi(x, u, \xi)\, \bar\xi(dx)\zeta(du). \tag{2.6.18}$$

Condition (2.6.16) is obviously more difficult to be verified than the corresponding conditions in *Theorems 2.3.2* and *2.4.2*. However, in a number of special cases it has a relatively simple form. For instance, for the *E*-criterion we have [cf. (2.1.7) and (2.1.14)]

$$\lambda_{\max} = \Psi(M, u) = \max_{u \in U} u^T M^{-1} u, . \tag{2.6.19}$$

where $U = \{u : u^T u = 1\}$. Some algebra shows that

$$\psi(x, u, \xi) = u^T M^{-1}(\xi) u - \left(u^T M^{-1}(\xi) f(x)\right)^2. \tag{2.6.20}$$

If a unique solution of (2.6.19) exists for the optimal design ξ^* then (2.6.16) is equivalent to the inequality

$$\left(f^T(x) u_{\max}\right)^2 \leq \lambda_{\max}^{-1}(\xi^*), \tag{2.6.21}$$

where $\lambda_{\max}$ is the largest eigenvalue of the matrix $M^{-1}(\xi^*) = D(\xi^*)$ or $\lambda_{\max}^{-1}$ is the smallest eigenvalue of $M(\xi^*)$; $u_{\max}$ is the corresponding eigenvector.

If (2.6.19) has more than one solution, then (2.6.21) must be replaced by the inequality

$$\sum_{\ell=1}^{\alpha'} \zeta_\ell (f^T(x) u_\ell)^2 \leq \lambda_{\max}^{-1}(\xi^*), \tag{2.6.22}$$

where α' is the multiplicity of $\lambda_{\max}(\xi^*)$ and $\{u_\ell\}_1^{\alpha'}$ is the set of the corresponding eigenvectors.

Intuitively we can say that maximizing the minimax criterion provides the experimenter with a design which guarantees reasonably good results even in the case of worst values of u. Unfortunately, sometimes, like in the above case, it happens that the worst u occurs more than once. Then we have to treat appropriately all these u's, and that explains the integration with respect to the measure ζ; in most cases, this integration is a weighted sum similar to (2.6.22).

The Response Function Contains Uncontrolled and Unknown Independent Variables. So far the independent variables $x \in X$ were assumed to be under control of the experimenter. Weather conditions in agricultural experiments, temperature, atmospheric pressure, wind direction in environmental studies, age, living conditions, gender in biological studies, etc. are examples for variables $u \in U$ that are out of the experimenters control. In the following, we split the independent variables into two groups and we replace $f(x)$ by $f(x, u)$, i.e.,

$$y = f^T(x, u)\theta + \varepsilon; \tag{2.6.23}$$

the variables $x \in X \subset R^k$ are assumed to be controlled whereas the variables $u \in U$ are assumed to be out of control. Then, the information matrix $M(\xi)$ becomes

$$M(\xi, u) = \int f(x, u) f^T(x, u) \xi(dx). \tag{2.6.24}$$

Of course, if u is known, (2.6.24) does not require any changes in the standard approach. Difficulties are faced when components of the vector u are unknown.

The corresponding design problem is very similar to what was considered in the previous subsection. We replace $\Psi[M(\xi), u]$ by $\Psi[M(\xi, u)]$ and add the assumption that all components of $f(x, u)$ are continuous with respect to u in U. For instance, for the D-criterion we have the following analogue of *Theorem 2.6.1*:

A necessary and sufficient condition for ξ^* to be optimal is the existence of the measure ζ^*, such that

$$\max_x d(x, \xi^*, \zeta^*) \le m \tag{2.6.25}$$

where

$$d(x, \xi, \zeta) = \int_{U(\xi)} f^T(x, u) M^{-1}(\xi, u) f(x, u) \, \zeta(du),$$

and

$$U(\xi) = \{u : \ u(\xi) = \arg \min_{u \in U} |M(\xi, u)|\}.$$

Verification of (2.6.25) is usually more difficult than verification of (2.6.16). We will use (2.6.25) and its analogues for other criteria in the case of nonlinear (with respect to θ) responses (cf. Section 5.6).

If the set U is compact, then Carathéodory's Theorem assures the existence of ζ^* [in (2.6.16) and (2.6.25)] with not more than $\dim U + 1$, where $\dim U$ means the dimension of U.

Chapter 3

Numerical Techniques

3.1 First Order Algorithm: D-criterion

All algorithms that we are going to discuss are iterative. Let ξ_s be the design that was obtained after $(s-1)$ iteration steps. At the s-th iteration step we hopefully will improve – as in all previous steps – the characteristics of the design. The new design, for instance, may be obtained according to

$$\xi_{s+1} = (1 - \alpha_s)\xi_s + \alpha_s\xi\,; \qquad (3.1.1)$$

this means that we reduce the number of observations which are taken in accordance with design ξ_s and that we, instead, take some observations at points which correspond to design ξ. How shall we choose those points or the design ξ?

Recall that

$$d(x, \xi_s) = f^T(x)D(\xi_s)f(x)$$

is the sensitivity function for the D-criterion (cf. *Table 2.1*) and that all support points of a D-optimal design coincide with maxima of the sensitivity function. Thus, we may expect that adding observations (or design measure) to the points where the function $d(x, \xi_s)$ achieves its maximum will improve the design ξ_s. The function $d(x, \xi_s)$ is the normalized variance of the estimated response at point x. In other words, in order to obtain an improved ξ_{s+1}, we have to move points of observation to those points where the variance of the response function is worst, i.e., has the largest value.

More formally, we can describe this idea in the following way:

(a) Given ξ_s, find

$$x_s = \arg\max_x d(x, \xi_s)\,. \qquad (3.1.2)$$

(b) Add point x_s to the design, i.e., construct

$$\xi_{s+1} = (1 - \alpha_s)\xi_s + \alpha_s\xi(x_s)\,,$$

where $\xi(x_s)$ is a unit measure atomized at x_s.

45

We will see how this step is repeatedly applied as the key element of an iterative procedure for the construction of optimal designs. The stopping rule of such procedures must be that we cannot find a point that does not belong to the support set of the constructed design and for which the sensitivity function is greater than anywhere else.

A "Forward" Iterative Procedure. Let us look for theoretical reasons why the iteration step (a-b) may be "good". First, we compare values of $|D(\xi_s)|$ and $|D(\xi_{s+1})|$ where ξ_{s+1} is defined by (3.1.1). As we have learned [cf. assumption (B4) from Section 2.3],

$$\log |D(\xi_{s+1})| = \log |D(\xi_s)| - \alpha \int d(x, \xi_s)\, \xi(dx) + o(\alpha, \xi_s, \xi)\,. \qquad (3.1.3)$$

If we aim at finding the design ξ with least value $\log |D(\xi)|$, minimization of the function $\log |D(\xi_{s+1})|$ with respect to the added measure $\alpha \xi(dx)$ is a natural thing to do. For sufficiently small α,

$$\min_{\xi} \log |D(\xi_{s+1})| \cong \min_{\xi} \left[\log |D(\xi_s)| - \alpha \int d(x, \xi_s)\, \xi(dx) \right]$$

$$= \log |D(\xi_s)| - \alpha \max_{\xi} \int d(x, \xi_s)\, \xi(dx)\,. \qquad (3.1.4)$$

Observing that

$$\max_{\xi} \int d(x, \xi_s)\, \xi(dx) = \max_{x} d(x, \xi_s)\,, \qquad (3.1.5)$$

we conclude that for small α the one-point design defined by (3.1.2) is the best choice. Note that (3.1.4) implies that the iteration step (a-b) improves the design by moving along the least directional derivative.

To apply our step (a-b) in an iterative procedure, we not only have to know the direction but also to choose the "step length" α_s. The following three choices for $\{\alpha_s\}$ are the most popular ones:

$$\lim_{s \to \infty} \alpha_s = 0, \quad \sum_{s=0}^{\infty} \alpha_s = \infty\,; \qquad (3.1.6)$$

$$\alpha_s = \arg \min_{\alpha} \Psi[(1 - \alpha)\xi_s + \alpha \xi(x_s)]\,; \qquad (3.1.7)$$

$$\alpha_s = \begin{cases} \alpha_{s-1}\,, & \text{if } \Psi[(1 - \alpha_{s-1})\xi_s + \alpha_{s-1}\xi(x_s)] < \Psi(\xi_s)\,, \\ \gamma \alpha_{s-1}\,, & \text{otherwise.} \end{cases} \qquad (3.1.8)$$

The quantity $\gamma < 1$ must be suitably chosen. Rule (3.1.7) results in the steepest descent and is thus called the the steepest descent rule.

Figure 3.1 illustrates the forward iterative procedure. For a polynomial of order 4 on $[-1, 1]$, a D-optimal design is to be derived. Starting point is the design $\xi_0 = \{\pm 1, \pm 0.5, 0; p_i = 0.2, \text{ for all } i\}$; in Figure 3.1, the corresponding sensitivity function is indicated by the non-bold line. The sensitivity functions for the design that results after 10 iterations and for the optimal design ($\xi^* = \{\pm 1, \pm 0.65, 0; p_i = 0.2, \text{ for all } i\}$) are drawn as the dashed and the bold line, respectively.

Figure 3.1: Sensitivity function for the D-criterion at three stages of the forward iterative procedure: for the initial design (non-bold line), an intermediate design (dashed line), and for the D-optimal design (bold line).

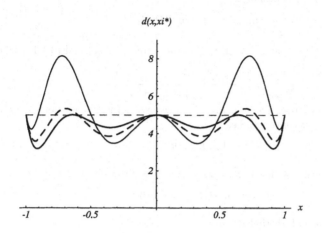

Using the relation

$$|A + BB^T| = |A| |I + B^T A^{-1} B| \qquad (3.1.9)$$

[cf. the result (B1) in the Appendix A], we can derive

$$
\begin{aligned}
|D^{-1}(\xi_{s+1})| &= |M(\xi_{s+1})| = |(1 - \alpha)M(\xi_s) + \alpha f(x_s)f^T(x_s)| \\
&= (1 - \alpha)^m \left(1 + \frac{\alpha d(x_s, \xi_s)}{1 - \alpha}\right) |M(\xi_s)|,
\end{aligned}
\qquad (3.1.10)
$$

and find for (3.1.7) that

$$\alpha_s = \frac{d(x_s, \xi_s) - m}{[d(x_s, \xi_s) - 1]m} . \qquad (3.1.11)$$

By means of

$$(A + BB^T)^{-1} = A^{-1} - A^{-1}B(I + B^T A^{-1} B)^{-1} B^T A^{-1} \qquad (3.1.12)$$

[cf. the result (B2) in the Appendix A], the following useful recursion can be derived:

$$D(\xi_{s+1}) = \frac{D(\xi_s)}{1 - \alpha} \left[I - \frac{\alpha f(x_s)f^T(x_s)D(\xi_s)}{1 - \alpha + \alpha d(x_s, \xi_s)}\right] . \qquad (3.1.13)$$

The iteration step (a-b) provides the "best" improvement. Does its repeated application lead to an optimal design when $s \to \infty$? The answer is positive. For instance, in the case of the steepest descent rule (3.1.7), the following theorem can be proved.

Theorem 3.1.1 *If ξ_0 is regular, then*

$$\lim_{s \to \infty} |D(\xi_s)| = \min_{\xi} |D(\xi)| . \tag{3.1.14}$$

Proof: Let $\Psi_s = |D(\xi_s)|$ and $\delta_s = d(x_s, \xi_s) - m$. From (3.1.10) and (3.1.11) follows that

$$\frac{\Psi_s}{\Psi_{s+1}} = \left(\frac{m + \delta_s}{m}\right)^m \left(\frac{m-1}{m-1+\delta_s}\right)^{m-1} . \tag{3.1.15}$$

By definition of our iterative step (a-b), the sequence $\{\Psi_s\}$ is nondecreasing and bounded; therefore it converges, i.e.,

$$\lim_{s \to \infty} \Psi_s = \Psi . \tag{3.1.16}$$

Let us assume that Ψ deviates by an amount $\gamma > 0$ from $\min_{\xi} |D(\xi)|$, i.e.,

$$\Psi - \min_{\xi} |D(\xi)| = \gamma > 0 . \tag{3.1.17}$$

Then from (2.3.16) follows that for all s

$$\delta_s \geq \gamma > 0 .$$

For $\delta_s > 0$, the ratio (3.1.15) is increasing:

$$\frac{\Psi_s}{\Psi_{s+1}} \geq 1 + \beta ,$$

where $\beta > 0$; consequently,

$$\lim_{s \to \infty} \Psi_s = -\infty$$

which contradicts (3.1.16). This proves the theorem.

Theorem 3.1.1 is valid not only for the steepest descent rule (3.1.7) but for all sequences $\{\alpha_s\}$ that were introduced above. For the rule (3.1.6) we have to modify part (b) of our iteration step (a-b):

(b') if $|D(\xi_s)| \leq K$, go to (a); K is usually selected as $K \geq |D(\xi_0)|$ where ξ_0 is the starting design; if $|D(\xi_{s+1})| > K$, set $\xi_{s+1} = \xi_0$ and proceed with (a).

Without this modification, (3.1.14) must be substituted by a weaker statement: $\lim_{s \to \infty} |D(\xi_s)|$ either is equal to $\min_{\xi} |D(\xi)|$ or approaches $-\infty$.

Real Numerical Procedures. Procedures based on (a-b), (3.1.2), and $\{\alpha_s\}$ defined by (3.1.6), (3.1.7), or (3.1.8) are a good introduction and help to understand the basic ideas and theory. However, they are hardly used for practical computing. It is difficult to describe all "tricks" that are used in algorithms for practical applications. In the following we describe some of them.

(1) The convergence rate is significantly improved if we, parallel to adding "good" points [cf. (3.1.2)], delete "bad" points from ξ_s: To do this we have to find

$$x_s^- = \arg \min_{x \in X_s} d(x, \xi_s), \qquad (3.1.18)$$

where $X_s = \text{supp}\,\xi_s$, and either to delete x_s^- from X_s or to reduce its weight. We may, for instance, replace (3.1.2) by

$$x_s = \arg \max \left[d(x_s^+, \xi_s) - m,\, m - d(x_s^-, \xi_s) \right], \qquad (3.1.19)$$

where x_s^+ is defined by (3.1.2), and use

$$\alpha_s = \begin{cases} \alpha_s', & \text{if } x_s = x_s^+, \\ -\min\left(\alpha_s', \frac{p_s}{(1-p_s)}\right), & \text{if } x_s = x_s^-; \end{cases} \qquad (3.1.20)$$

α_s' is given by one of the rules (3.1.6), (3.1.7), or (3.1.8), and p_s is the weight of point x_s^- in the design ξ_s.

(2) It is computationaly cumbersome but not necessary to solve (3.1.2) precisely. It is sufficient to determine x_s such that $d(x_s, \xi_s) - m \geq \delta_s > 0$.

(3) The points x_s have the tendency to cluster around support points of the optimal design. We can avoid that in the following way: Check whether x_s fulfills the inequality

$$\max_{x_{is} \in X_s} (x_{is} - x_s)^T (x_{is} - x_s) \leq r, \qquad (3.1.21)$$

where the constant r is chosen on practical grounds. If an x_{is} is close enough to x_s to satisfy (3.1.21) then x_{is} is to be deleted and its weight to be added to that of x_s. In older publications it has been recommended to use their weighted average instead of x_s and x_{is}. We found that our rule works better and is, in addition, in better agreement with theory: Observations have to be taken at points where the function $d(x, \xi^*)$ achieves its maxima. Obviously, $d(x, \xi_s)$ becomes "closer" to $d(x, \xi^*)$ if $s \to \infty$. Therefore, for large s, the point x_s is closer to the corresponding support point of an optimal design than neighboring points that were found for smaller s.

(4) In (1) the choice between adding or deleting is controlled by (3.1.19). Sometimes it is more convenient first to make n^+ adding steps with x_s from (3.1.2) and then to proceed with n^- deleting steps based on (3.1.18). The number n^+ is called the length of the forward excursion, and n^- is called the length of the backward excursion.

(5) Usually the iterative procedure is considered to be completed when some stopping rule is fulfilled. From the convexity of the function $\log |D(\xi)|$ follows that

$$\log \frac{|D(\xi)|}{|D(\xi^*)|} \leq \max_x d(x, \xi) - m \qquad (3.1.22)$$

[cf. (2.3.16)]. At every iteration, $\max_x d(x, \xi_s)$ is computed and, therefore, it is not difficult to know how close $|D(\xi_s)|$ is to $|D(\xi^*)|$ and to formulate an appropriate stopping rule.

3.2   First Order Algorithm: The General Case

In what follows we presume that assumptions (A1)-(A2) and (B1)-(B4) hold and that we, therefore, can use *Theorem 2.3.2*. However, we modify (B4):

(B4′) For any $\xi \in \Xi(q)$ and $\overline{\xi} \in \Xi$

$$\Psi[(1 - \alpha)M(\xi) + \alpha M(\overline{\xi})]$$
$$= \ \Psi\left[M(\xi)\right] + \alpha \int \psi(x, \xi)\, \overline{\xi}(dx) + O(\alpha^2|\xi, \overline{\xi}), \tag{3.2.1}$$

where $O(\alpha^2|\xi, \overline{\xi}) \le \alpha^2 K_q$ for some $K_q > 0$.

This means that the first order approximation is valid uniformly with respect to ξ and $\overline{\xi}$.

Applying the ideas of Section 3.1, we consider the updating relation

$$\xi_{s+1} = (1 - \alpha_s)\xi_s + \alpha\xi^s \tag{3.2.2}$$

where

$$\xi^s = \arg\min_\xi \Psi[(1 - \alpha_s)\xi_s + \alpha_s\xi] \,. \tag{3.2.3}$$

We can replace (3.2.3) by the first order approximation

$$\xi^s = \arg\min_\xi \left\{ \Psi\left[M(\xi_s)\right] + \alpha_s \int \psi(x, \xi_s)\, \xi(dx) \right\} \,; \tag{3.2.4}$$

the corresponding algorithm is called the "first order algorithm". Obviously, the iterative procedures considered in Section 3.1 are special cases of the first order algorithm.

The simplicity and popularity of first order algorithms is explained by the fact that the design ξ^s does not depend on α_s and that

$$\xi^s = \arg\min_\xi \int \psi(x, \xi_s)\xi(dx) = \arg\max_\xi \int \phi(x, \xi_s)\xi(dx) \tag{3.2.5}$$

[cf. the definition of $\phi(x, \xi)$ in Section 2.4]. The solution of (3.2.5) is a design measure $\xi(x_s)$ that is atomized at the point

$$x_s = \arg\min_x \psi(x, \xi_s) = \arg\max_x \phi(x, \xi_s). \tag{3.2.6}$$

This fact is another motivation why the function $\phi(x, \xi)$ is called the sensitivity function and why it is tabulated for the various optimality criteria (see *Table 2.1*).

In Section 3.1 we discuss iterative algorithms that are based on the D-criterion. Similarly, (3.2.2) together with (3.2.5) and (3.2.6) allows to develop a number of first order algorithms, all of which can be embedded in the following scheme:

(a) Given $\xi_s \in \Xi_q$ find

$$x_s = \arg\min_x \psi(x, \xi_s) \,.$$

(b) Choose α_s from $0 \leq \alpha_s \leq 1$ and construct

$$\xi_{s+1} = (1 - \alpha_s)\xi_s + \alpha_s \xi(x_s).$$

A sequence $\{\alpha_s\}$ may be chosen, for instance, according to (3.1.6), (3.1.7), or (3.1.8).

Theorem 3.2.1 *If the assumptions of* Theorem 2.3.2 *and (B4') hold, then for* $\{\alpha_s\}$ *defined by (3.1.7)*

$$\lim_{s \to \infty} \Psi(\xi_s) = \Psi^* = \min_{\xi} \Psi(\xi). \tag{3.2.7}$$

Proof: By definition, $\{\Psi(\xi_s)\}$ is decreasing and, therefore, converges:

$$\lim_{s \to \infty} \Psi(\xi_s) = \Psi_\infty. \tag{3.2.8}$$

Assume that Ψ_∞ deviates from the optimal value Ψ^*; then

$$\Psi_\infty - \Psi^* = \delta > 0. \tag{3.2.9}$$

From the convexity of $\Psi(\xi)$ follows (cf. also *Theorem 2.3.2*) that

$$\min_x \psi(x, \xi_s) \leq \int \psi(x, \xi_s)\xi^*(dx) \leq \Psi(\xi^*) - \Psi(\xi_s) \leq -\delta. \tag{3.2.10}$$

Rule (3.1.7) together with assumption (B4') and (3.2.10) imply that

$$\Psi(\xi_s) - \Psi(\xi_{s+1}) \geq \delta/4K_q > 0$$

and, consequently,

$$\lim_{s \to \infty} \Psi(\xi_s) = -\infty$$

which contradicts (3.2.8); this proves the theorem.

The comments concerning the sequences $\{\alpha_s\}$ that are made after the proof of *Theorem 3.1.1* apply again: Under the conditions mentioned there, *Theorem 3.2.1* is valid for all three choices (3.1.6), (3.1.7), and (3.1.8).

Similar to the case treated in Section 3.1, a significant improvement in the rate of convergence is reached if we admit the possibility to reallocate weights within the design ξ_s, i.e., if

$$x_s = \arg \min \left[\psi(x_s^+, \xi_s), -\psi(x_s^-, \xi_s) \right], \tag{3.2.11}$$

where $x_s^+ = \arg \min_{x \in X} \psi(x, \xi_s)$ and $x^- = \arg \max_{x \in X} \psi(x, \xi_s)$. In this case, $\{\alpha_s\}$ as defined in (3.1.20) are to be applied.

All modification discussed in the concluding part of Section 3.1 are valid here and help to make the proposed algorithm more practical.

Regularization of the Numerical Procedure. If the rather restrictive assumption (B4') is not valid, sequences $\{\alpha_s\}$ that do not guarantee the monotonicity of $\{\Psi(\xi_s)\}$ may cause not-converging iterative procedures. This can happen, for instance, if the optimal design is singular. Of course, to get out of the problem, we can try various sequences of $\{\alpha_s\}$ and various starting designs ξ_0; this often is successful.

One of the most reliable and still relatively simple approaches is based on the concept of regularization as introduced in Section 2.6: We replace the minimization of $\Psi(\xi)$ by the minimization of $\Psi_\gamma(\xi) = \Psi\left[(1-\gamma)M(\xi) + \gamma M(\xi_0)\right]$, where ξ_0 is some regular design [cf. (2.6.1)]. If γ and ξ_0 are fixed, the changes in the iterative procedure are obvious. One has to modify the function $\psi(x,\xi)$ (or $\phi(x,\xi)$). For instance, for the D-criterion we set $\psi(x,\xi)$ to be

$$
\begin{aligned}
\psi(x,\xi) &= (1-\gamma)\left[\operatorname{tr} M^{-1}(\overline{\xi})M(\xi) - d(x,\overline{\xi})\right] \\
&= (1-\gamma)\left[\int d(x,\overline{\xi})\xi(dx) - d(x,\overline{\xi})\right],
\end{aligned}
\tag{3.2.12}
$$

where $\overline{\xi} = (1-\gamma)\xi + \gamma\xi_0$. For the linear criterion with loss matrix A we get

$$
\begin{aligned}
\psi(x,\xi) &= (1-\gamma)[\operatorname{tr} AM^{-1}(\overline{\xi})M(\xi)M^{-1}(\overline{\xi}) \\
&\quad - f^T(x)M^{-1}(\overline{\xi})AM^{-1}(\overline{\xi})f(x)] \\
&= (1-\gamma)\left[\int \phi(x,\overline{\xi})\xi(dx) - \phi(x,\overline{\xi})\right].
\end{aligned}
\tag{3.2.13}
$$

The last line is valid for any criterion that satisfies (B4′).

From the theoretical point of view, it makes sense to diminish γ for increasing s. However, the rate of decrease must be significantly less than that for α_s. We may expect that for an appropriate choice of the sequence $\gamma(s)$ with $\gamma(s) \to 0$

$$
\lim_{s\to\infty} \Psi(\xi_{\gamma(s),s}) = \min_\xi \Psi(\xi).
\tag{3.2.14}
$$

Second Order Algorithm. The main assumption that is needed in order to develop a first order iterative procedure is the existence of the first directional derivative [cf. (B4)]. Evidently, the idea can be extended in a straight manner if higher derivatives exist. The corresponding algorithms have a better rate of convergence, but are more complicated. Thus, we face the usual trade-off between efficiency and complexity of the algorithm. We shortly discuss in the following a second order iterative procedure for the D-criterion. This might serve as a motivation for the curious reader to try generalizations for higher order procedures and other criteria.

We start again with the updating relation

$$
\xi_{s+1} = (1-\alpha)\xi_s + \alpha\xi.
\tag{3.2.15}
$$

Some algebra shows that

$$
\begin{aligned}
\log|D(\xi_{s+1})| &= \log|D(\xi_s)| + \alpha\int(m - d(x,\xi_s))\xi(dx) \\
&\quad + \frac{\alpha^2}{2}\int\int[d^2(x,x',\xi_s) - d(x,\xi_s) - d(x',\xi_s) + m]\xi(dx)\xi(dx') \\
&\quad + O(\alpha^2,\xi,\overline{\xi}).
\end{aligned}
\tag{3.2.16}
$$

It is not difficult to prove the existence of a two-point design $\xi_s = (p_1, x_1, p_2, x_2)$ that maximizes the decrement at the s-th iteration. The weights p_1 and p_2 may be found analytically for given x_1 and x_2. Therefore, at every iteration we have to solve an optimization problem in $X \otimes X$. Instead of a simple one-point correction as described by (3.2.15) we have to use

$$\xi_{s+1} = (1 - \alpha_1 - \alpha_2)\xi_s + \alpha_1\xi(x_{1s}) + \alpha_2\xi(x_{2s}) \tag{3.2.17}$$

where $\alpha_j = \alpha p_j$, $j = 1, 2$. Note that the recursive formulas (3.1.10) and (3.1.14) are not applicable here.

3.3 Finite Sample Size

Rounding Procedures. Since we have introduced the design measure concept in Section 2.2, we have neglected the fact that for real experiments the number r_i of observations to be taken at each support point x_i is integer and, consequently, the weights $p_i = r_i/N$ are discrete. When it is necessary to emphasize the discrete character of weights we call the corresponding designs "discrete" ; other usual names are "exact" and "finite sample size" design.

If for a continuous (approximate) optimal design ξ^* the total number N of available observations is much greater than the number n of support points, we can achieve a satisfactory result using the standard numerical rounding:

$$r_i^* = \begin{cases} [p_iN]^+, & \text{if } [p_iN]^+ - p_iN \leq 0.5, \\ [p_iN]^-, & \text{otherwise,} \end{cases} \tag{3.3.1}$$

where $[u]^+$ is the least integer that is greater than u, and $[u]^-$ is the greatest integer that is less than u. Occasionally, (3.3.1) has to be violated to keep $\sum_{i=1}^n r_i^* = N$.

Let ξ_N^* be a discrete optimal design, i.e., an exact solution of $\min_{\xi_N} \Psi[M(\xi_N)]$ [cf. (2.1.4)]; and let $\tilde{\xi}_N$ be a design that is a rounded version of ξ^*; the index N indicates the sample size. Of course, it is of crucial interest to know the difference $\Psi(\tilde{\xi}_N) - \Psi(\xi_N^*)$. The following result may help to evaluate this difference in various settings.

Theorem 3.3.1 *Let n be the number of support points in ξ^* and let function $\Psi(M)$ be monotonic and homogeneous, i.e., (2.2.1) and (2.2.2) hold, then*

$$\Psi[M(\xi^*)] \leq \Psi[M(\xi_N^*)] \leq \frac{\gamma(N-n)}{\gamma(N)}\Psi[M(\xi^*)]. \tag{3.3.2}$$

Proof: From (2.2.1) and (2.2.2) follows that

$$\gamma(N-n)\Psi[M(\xi^*)] = \Psi[(N-n)M(\xi^*)]$$
$$\geq \Psi\left[\sum_{i=1}^n [Np_i^*]^- f(x_i^*)f^T(x_i^*)\right]$$
$$\geq \Psi[NM(\xi_N^*)] \geq \Psi[NM(\xi^*)] = \gamma(N)\Psi[M(\xi^*)].$$

The above inequalities obviously lead to (3.3.2).

The upper bound (3.3.2) can be lowered for some criteria. E.g., for the D-criterion we get from (3.3.2) that

$$\frac{|D(\xi_N^*)|}{|D(\xi^*)|} \leq \left(\frac{N}{N-n}\right)^m.$$

At the same time it is known that

$$\frac{|D(\xi_N^*)|}{|D(\xi^*)|} \leq \frac{N^m}{N(N-1)\dots(N+1-m)}.$$

Let us note that there are lucky cases where we have not to care for rounding procedures. For instance, from Section 2.5 we know that for the trigonometric regression on the interval $[0, 2\pi)$ any design with equidistant support points $(n \geq m)$ and equal weights is D-optimal. Therefore, for any $N > m$ we can easily construct the exact D-optimal design. Allocate at each of N equidistant points just one observation. A similar situation occurs for the first order polynomial regression on the sphere. Allocation of the observations at N "equidistant" points gives the exact D-optimal design. Of course, for some N and some dimensions this is not an easy problem.

Necessary Conditions. Most necessary conditions are based on a simple idea. A distortion of an optimal design cannot lead to an improvement. For instance, if an observation is moved from the support point x to x', some exercise in matrix algebra [cf. the result (B2) of the Appendix A] leads to the following formula for the D-criterion:

$$\frac{|D(\xi_{old})|}{|D(\xi_{new})|} = 1 - N^{-1}[d(x', \xi_{old}) - d(x, \xi_{old})]$$

$$- N^{-2}[d(x, \xi_{old})d(x', \xi_{old}) - d^2(x, x', \xi_{old})]. \qquad (3.3.3)$$

From this, we may conclude the following result.

Theorem 3.3.2 *For any point $x_i^* \in \operatorname{supp} \xi_N^*$ and any $x \in X$*

$$d(x_i^*, \xi_N^*) \geq d(x, \xi_N^*) - N^{-1}[d(x_i^*, \xi_N^*)d(x, \xi_N^*) - d^2(x_i^*, x, \xi_N^*)]. \qquad (3.3.4)$$

Summation of both sides of (3.3.4) with respect to x_i^* and the fact that for any design ξ

$$\sum_{i=1}^n p_i d(x_i, \xi) = \sum_{i=1}^n p_i f^T(x_i) D(\xi) f(x_i)$$

$$= \operatorname{tr} D(\xi) \sum_{i=1}^n p_i f^T(x_i) f(x_i) = \operatorname{tr} D(\xi) M(\xi) = m \qquad (3.3.5)$$

and

$$\sum_{i=1}^n p_i d^2(x_i, x, \xi) = \sum_{i=1}^n p_i f^T(x) D(\xi) f(x_i) f^T(x_i) D(\xi) f(x)$$

$$= f^T(x) D(\xi) \sum_{i=1}^n p_i f^T(x_i) f^T(x_i) D(\xi) f(x)$$

$$= f^T(x) D(\xi) f(x) = d(x, \xi) \qquad (3.3.6)$$

lead us to the inequality

$$d(x, \xi_n^*) \leq m \frac{N}{N + 1 - m} \, ; \tag{3.3.7}$$

compare this inequality with the last part of *Theorem 2.4.1* for $N \to \infty$.

Similar results may be derived for other optimality criteria. E.g., for the linear criterion the analogue for (3.3.4) from *Theorem 3.3.2* is

$$[1 + N^{-1} d(x, \xi_N^*)] \phi(x_i^*, \xi_N^*) \tag{3.3.8}$$
$$\geq \left[1 - N^{-1} d(x, \xi_N^*)\right] \phi(x, \xi_N^*) + 2N^{-1} d(x_i^*, x, \xi_N^*) \phi(x_i^*, x, \xi_N) \, ,$$

where ξ_N^* is assumed to be regular.

Numerical Search. Not much has been done in the development of special algorithms for discrete designs. Almost all existing and widely used algorithms are modifications of algorithms discussed in Section 3.2. For instance, if we choose the initial design ξ_0 of a procedure that is based on the general iteration step (a-b) in Section 3.2 such that $p_i = r_i/N$, and if we set $\alpha_s \equiv N^{-1}$ for all s, we come to the family of the first order algorithms for the construction of optimal discrete designs.

If the iterations with $\alpha_s \equiv N^{-1}$ do not lead to further improvement after some s, the computations may be continued with a diminishing α_s in order to evaluate how much the so obtained discrete design deviates from a continuous optimal design ξ^*. Generally, the iterative procedure results in a local solution. To assure global optimality, it is recommended to make several attempts to find the optimal design, starting with different inital designs.

Another family of algorithms can be constructed by using formulas similar to (3.3.4). E.g., for the D-criterion an iterative procedure can be constructed where pairs of points (x_s^+, x_s^-) are determined according to

$$(x_s^+, x_s^-) = \arg \max_{x \in X} \{ \max_{x_i \in X_s} d(x, \xi_s) - d(x_i, \xi_s)$$
$$- N^{-1}[d(x, \xi_s) d(x_i, \xi_s) - d^2(x, x_i, \xi_s)] \} \, . \tag{3.3.9}$$

In each iteration step, the design is improved by deleting x_s^- and adding x_s^+. This algorithm is computationally more demanding than first order algorithms but generally leads to better results.

Chapter 4

Optimal Design under Constraints

4.1 Cost Constraints

In Section 2.3, an optimal design ξ^* is defined as

$$\xi^* = \arg\min_{\xi} \Psi[M(\xi)]; \qquad (4.1.1)$$

it is assumed and extensively used that

$$\int \xi(dx) = 1. \qquad (4.1.2)$$

Actually, (4.1.2) may be considered as a continuous analogue of the constraint that is imposed on the number of observations, viz.,

$$\sum_{i=1}^{n} r_i = N. \qquad (4.1.3)$$

Constraint (4.1.2) or (4.1.3) may be replaced by a constraint that is related to the costs of the experiment. For instance, if the cost of one observation amounts, depending on x, to $\underline{\zeta}(x)$ and the total cost is limited by c, then we come to constraints

$$N^{-1}\sum_{i=1}^{n} r_i \, \underline{\zeta}(x_i) \leq c \ \text{ or } \ \int \underline{\zeta}(x)\,\xi(dx) \leq c. \qquad (4.1.4)$$

If this constraint is the only one that must be satisfied, the transformations

$$\xi'(dx) = c^{-1}\underline{\zeta}(x)\,\xi(dx) \ \text{ and } \ f'(x) = \sqrt{c\underline{\zeta}^{-1}(x)}\,f(x) \qquad (4.1.5)$$

bring us back to the standard case (4.1.1) and (4.1.2).

In the following, we will use $\zeta(x) = \underline{\zeta}(x) - c$ rather than $\underline{\zeta}(x)$. This allows us to rewrite (4.1.4) in the form

$$\Phi(\xi) = \int \zeta(x)\,\xi(dx) \leq 0. \qquad (4.1.6)$$

The structure of constraint (4.1.6) covers a wide range of cases and comprises the so-called "linear constraint". If ℓ constraints have to be taken into account, the vector-function ζ contains ℓ components.

As already mentioned, one constraint alone such as (4.1.2) or (4.1.4) does not make it necessary that we rephrase our optimization problem as shown above. However, the situation changes if an experiment is constrained both in the number of observations and in the costs. In this case, we have to consider the more general problem

$$\xi^* = \arg \max_\xi \Psi[M(\xi)] \tag{4.1.7}$$

$$\text{s.t. } \int \xi(dx) = 1 \text{ and } \int \zeta(x)\xi(dx) \le 0.$$

The transformations (4.1.5) do not help us to solve this problem. Note that (4.1.7) is a general formulation as we allow $\zeta(x)$ to have $\ell \ge 1$ components.

Constrained Optimal Design. The properties of the constrained optimal design (4.1.7) are summarized in the following theorem.

Theorem 4.1.1 *Let assumptions (A1)-(A2) and (B1)-(B4) from Section 2.3 hold; let us further assume that, in addition to condition (B4), the designs from $\Xi(q)$ satisfy (4.1.6) and that*

(C) $\zeta(x)$ is continuous for all $x \in X$.

Then the following statements hold.

1. *For any optimal design there exists a design with the same information matrix that contains not more than $m(m + 1)/2 + \ell$ support points.*

2. *A necessary and sufficient condition for a design ξ^* to be optimal is that it fulfills the inequality*

$$\min_x q(x, u^*, \xi^*) \ge 0,$$

where $q(x, u, \xi) = \psi(x, \xi) + u^T \zeta(x)$, $u^* = \arg \max_{u \in U} \min_x q(x, u, \xi^*)$, and $U = \{u : u \in R^\ell, u_i \ge 0, i = 1, \ldots, \ell\}$.

3. *The set of optimal designs is convex.*

4. *The function $q(x, u^*, \xi^*)$ achieves zero almost everywhere in supp ξ^*.*

Proof: 1. Note that any pair $(M(\xi), \Phi(\xi))$ belongs to the convex hull of

$$\{m(x), \zeta(x)\} \in R^{m(m+1)/2+\ell}$$

and apply Carathéodory's Theorem (cf. the proof of *Theorem 2.3.1*).

2. Add the constraint

$$\int \zeta(x)\,\xi(dx) \le 0 \tag{4.1.8}$$

to the inequality

$$\min_{\xi} \int \psi(x, \xi^*)\, \xi(dx) \geq 0 \,. \tag{4.1.9}$$

Inequality (4.1.9) together with (4.1.8) constitutes a necessary and sufficient condition of optimality of ξ^*. In general, no single point design exists [cf. (2.3.13)] that satisfies (4.1.9) and (4.1.8). The Lagrangian theory [see, e.g., Laurent (1972), Chpt. 7] indicates the duality of the optimization problem (4.1.9) and (4.1.8) and the maximin problem

$$\max_{u \in U} \min_{\xi} \int q(x, u, \xi^*)\, \xi(dx) \,,$$

or equivalently

$$\max_{u \in U} \min_{x} q(x, u, \xi^*) \,,$$

confirming the assertion of the theorem. The proofs of the other two parts of the theorem coincide with that of the corresponding results for the standard case.

Example: Polynomial Regression. Let us consider the design problem for a one-dimensional polynomial response function with $f_i(x) = x^{i-1}$, $i = 1, \ldots, m$, and $|x| \leq 1$, and the D-criterion $\Psi(M) = -\log|M|$ with linear constraints

$$\int_{-1}^{1} \zeta(x)\, \xi(dx) \leq 0 \,.$$

Let $\{f, \zeta\}$ constitute a Tchebysheff system on $|x| \leq 1$. The function

$$q(x, u, \xi) = m - \sum M_{ij}^{-1} x^{i-1} x^{j-1} + u^T \zeta(x) \,,$$

where M_{ij}^{-1} are the elements of matrix M^{-1}, is a linear combination of $2m + \ell$ Tchebysheff functions with nonzero coefficients. Therefore [see, e.g., Karlin and Studden (1966)], this function has not more than $2m + \ell$ roots and, consequently, has not more than $m + \ell/2$ (if ℓ is even) or $m + (\ell+1)/2$ (if ℓ is odd) minima on the interval $|x| \leq 1$. But, according to *Theorem 4.1.1*, the minima have to coincide with the support points. In this example, the number of support points is considerably less than $m(m+1)/2 + \ell$.

Numerical Procedures. The numerical procedures discussed in Chapter 3 may be modified in order to handle the optimization problem defined in (4.1.7). The simplest version of an iterative procedure is based on the first order algorithm and comprises the following steps:

(a) Given $\xi_s \in \Xi(q)$, find

$$\xi^s = \arg\min_{\xi} \int \psi(x, \xi_s)\xi(dx) \quad \text{s.t.} \quad \int \zeta(x)\xi(dx) \leq 0 \,. \tag{4.1.10}$$

(b) Choose $0 \leq \alpha_s \leq 1$ and construct

$$\xi_{s+1} = (1 - \alpha_s)\xi_s + \alpha_s \xi^s \,. \tag{4.1.11}$$

Unlike in the standard case, (4.1.10) cannot be reduced to the minimization of the function $\psi(x, \xi_s)$ which is a relatively simple problem. The minimization again can be simplified by applying Carathéodory's Theorem: We can prove the existence of a design ξ^s that has not more than $\ell + 1$ support points. Then, (4.1.10) can be reduced to a finite-dimensional problem:

$$\xi^s = \left\{ \begin{matrix} x_1^s & \cdots & x_{\ell+1}^s \\ p_1^s & \cdots & p_{\ell+1}^s \end{matrix} \right\} = \arg \min_{\{p_j, x_j\}} \sum_{j=1}^{\ell+1} p_j \psi(x_j, \xi_s)$$

$$\text{s.t.} \quad \sum_{j=1}^{\ell+1} p_j \zeta(x_j) \leq 0, \quad \sum_{j=1}^{\ell+1} p_j = 1.$$

The proof of convergence of this iterative procedure is identical to that for the standard case; see *Theorem 3.2.1*.

This problem is practically solvable if the number of constraints ℓ and the dimension of X are reasonably small.

Lagrangian Approach. On an intuitive level it is clear that by minimizing the function

$$\Upsilon(\xi) = \Psi(\xi) + \lambda^T c(\xi) \tag{4.1.12}$$

with $\lambda_i > 0$, $i = 1, \ldots, \ell$, and $c(\xi) = \int \zeta(x) \xi(dx)$, we can find a design ξ^* that is a solution of the optimization problem (4.1.7) with $c_\lambda = c(\xi_\lambda^*)$. We return to the cost function $\underline{\zeta}(x) = \zeta(x) + c$ because it may help to a better understanding of this approach.

Theorem 4.1.2 *If the assumptions of* Theorem 4.1.1 *hold, then the design*

$$\xi_\lambda = \arg \min_\xi [\Upsilon(\xi)] \tag{4.1.13}$$

is a solution of the constrained optimization problem (4.1.7) with $c_\lambda - c(\xi_\lambda)$.

Proof: Obviously, $\Upsilon(\xi) = \Psi(\xi) + \lambda^T c(\xi)$ satisfies all assumptions of *Theorem 2.3.2*. According to this theorem, ξ_λ is optimal if and only if

$$\min_x \left[\psi(x, \xi_\lambda) + \lambda^T \underline{\zeta}(x) \right] \geq 0. \tag{4.1.14}$$

But $\psi(x, \xi) + \lambda^T \underline{\zeta}(x)$ coincides with $q(x, u, \xi)$ in *Theorem 4.1.1* if we set $u = \lambda$ and $\zeta_\lambda(x) = \underline{\zeta}(x) - c_\lambda$. Furthermore,

$$\min_x \max_u \left[\psi(x, \xi_\lambda) + u^T \zeta_\lambda(x) \right] \geq \min_x \left[\psi(x, \xi_\lambda) + \lambda^T \zeta_\lambda(x) \right] \geq 0.$$

According to *Theorem 4.1.1* this inequality is sufficient to assert that

$$\xi_\lambda = \arg \min_\xi \Psi(\xi) \quad \text{s.t.} \quad \int \zeta_\lambda(x) \xi(dx) \leq 0; \tag{4.1.15}$$

this completes the proof.

Theorem 4.1.2 tells us that ξ_λ is a solution of the constrained problem which coincides with the original problem (4.1.7) if $c - c_\lambda = 0$. The solution of (4.1.13) can be found by means of the numerical procedures developed in Chapter 3. To apply them we need to guess values for λ such that c is close to c_λ. As a measure of closeness we may, e.g., introduce

$$\beta(\lambda) = (c - c_\lambda)^T A (c - c_\lambda), \tag{4.1.16}$$

where the positive definite matrix A reflects the significance of each of the ℓ constraints. Consequently, we have to consider the "empirical" optimization problem

$$\lambda^* = \arg\min_\lambda \beta(\lambda) \quad \text{s.t.} \quad \lambda_i \geq 0, \ i = 1,\dots,\ell. \tag{4.1.17}$$

Assuming that it is not difficult to find ξ_λ and c_λ for a given λ, we can use any numerical procedure to find a solution of (4.1.17).

4.2 Constraints for Auxiliary Criteria

The approach developed in Section 4.1 can be applied with some modifications to a more general design problem:

$$\xi^* = \arg\min_\xi \Psi(\xi) \quad \text{s.t.} \quad \Phi(\xi) \leq 0, \tag{4.2.1}$$

where the components of $\Phi^T = (\Phi_1,\dots,\Phi_\ell)$ can be from a very general class of functions of ξ [see (C1)-(C3) below]. All constraints are assumed to be active for ξ^*. Let us assume in addition to assumptions (A1)-(A2) and (B1)-(B4) from Section 2.3 that

(C1) all components of $\Phi(\xi)$ are convex;

(C2) there exists a real number q such that the set

$$\{\xi \colon \Psi(\xi) \leq q < \infty, \ \Phi(\xi) \leq 0\} = \Xi(q) \tag{4.2.2}$$

is not empty;

(C3) for any $\xi \in \Xi(q)$ and $\bar\xi \in \Xi$,

$$\Phi\left[(1 - \alpha)\xi + \alpha\bar\xi\right] = \Phi(\xi) + \alpha \int \zeta(x,\xi)\,\bar\xi(dx) + o(\alpha|\xi,\bar\xi). \tag{4.2.3}$$

Assumptions (C1), (C2), and (C3) are almost identical to (B1), (B3), and (B4), respectively, but are concerned with $\Phi(\xi)$.

Extended version of Theorem 4.1.1. The analysis of (4.2.1) is based on ideas of Section 4.1 and makes use of the linearization or first order approximation of $\Phi(\xi)$ near the optimal design.

For instance, a necessary and sufficient condition of optimality of the design ξ^* is identical to the one stated in part 2 of *Theorem 4.1.1* with obvious replacement of $\zeta(x)$ by $\zeta(x, \xi^*)$. In this case we will speak of the "extended version of *Theorem 4.1.1*".

Cases that can be embedded in the design problem (4.2.1) are of the following type: Let the main goal of an experiment be to minimize a certain criterion, while at the same time some auxiliary criteria should have acceptable values. Typical situations are uncertainty (a) with respect to the model and (b) with respect to the optimality criterion to be applied.

Uncertainty Concerning the Model. To get some feeling of how the extended version of *Theorem 4.1.1* works, let the design problem be to find a D-optimal design for the response $\theta_0^T f_0(x)$ while ensuring that this design is still reasonably "good" for the alternative responses $\theta_j^T f_j(x)$, $j = 1, \ldots, \ell$, i.e.,

$$
\begin{aligned}
\Psi(\xi) &= -\log|M_0(\xi)|, &\text{(4.2.4)} \\
\Phi_j(\xi) &= -\log|M_j(\xi)| - c_j, \quad j = 0, \ldots, \ell, &\text{(4.2.5)}
\end{aligned}
$$

where $M_j(\xi) = \int f_j(x) f_j^T(x)\, \xi(dx)$. Recall that for the D-criterion

$$
\begin{aligned}
\psi(x, \xi) &= m_0 - d_0(x, \xi), \\
\zeta_j(x, \xi) &= m_j - d_j(x, \xi),
\end{aligned}
$$

where m_j is the number of unknown parameters in the j-th response and $d_j(x, \xi) = f_j^T(x) M_j^{-1}(\xi) f_j(x)$. Assume further that all basis functions are continuous in X and that assumption (C2) holds. Then we can verify the validity of all assumptions that are needed for the extended version of *Theorem 4.1.1* to be true.

From this theorem follows that a necessary and sufficient condition of optimality of ξ^* is the existence of u^* such that

$$
d(x, \xi^*) + \sum_{j=1}^{\ell} u_j^* d_j(x, \xi^*) \leq m_0 + \sum_{j=1}^{\ell} u_j^* m_j, \qquad \text{(4.2.6)}
$$

while $\Phi(\xi^*) = 0$. The equality holds at all support points of ξ^*.

Uncertainty Concerning the Optimality Criterion. Above, we have tried to achieve the least determinant of the dispersion matrix for the response $\theta_0^T f_0(x)$ while the corresponding determinants for competing models are relatively small. In applications we may meet another type of problem that, however, can be dealt with in a very similar way.

Let $\theta^T f(x)$ be the model of interest. We aim at finding a design that minimizes $-\log|M(\xi)|$, but keeps, e.g., $\operatorname{tr} M^{-1}(\xi)$ at relatively low level. This means that

$$
\Psi(\xi) = -\log|M(\xi)| \qquad \text{(4.2.7)}
$$

and

$$
\Phi(\xi) = \operatorname{tr} M^{-1}(\xi) - c. \qquad \text{(4.2.8)}
$$

Similar to the previous case, the extended version of *Theorem 4.1.1* tells us that a necessary and sufficient condition of optimality of ξ^* is the existence of $u^* \geq 0$ such that

$$d(x, \xi^*) + u^* f^T(x) M^{-2}(\xi^*) f(x) \leq m + u^* \operatorname{tr} M^{-1}(\xi^*). \tag{4.2.9}$$

The reader might try to formulate combinations of the above two types of design problems to match practical needs.

4.3 Directly Constrained Design Measures

In many spatial experiments, two very closely sited sensors do not give much more information than each one of them. In other words the density of sensor allocation must be lower than some reasonable level. For a design measure $\xi(dx)$ this means that

$$\xi(dx) \leq \omega(dx),$$

where $\omega(dx)$ describes the maximal possible "number" of sensors per dx, and it is assumed that $\int \omega(dx) \geq 1$. Of course, if the equality holds we cannot do anything better than assume that $\xi(dx) = \omega(dx)$. Otherwise the following optimization problem must be considered:

$$\xi^* = \arg \min \Psi(\xi) \quad \text{s.t.} \quad \xi(dx) \leq \omega(dx). \tag{4.3.1}$$

The design ξ^* can be called a (Ψ, ω)-optimal design. Whenever this is not ambiguous we will skip (Ψ, ω).

Le us assume in addition to (B1)-(B4) from Section 2.3 that

(D) $\omega(dx)$ is atomless, i.e., for any ΔX exists a subset $\Delta X' \subset \Delta X$ such that [cf. Karlin and Studden (1966), Chpt. VIII.12]

$$\int_{\Delta X'} \omega(dx) < \int_{\Delta X} \omega(dx).$$

The sets Ξ and $\Xi(q)$ in (B4) have to satisfy the constraint of (4.3.1). Let Ξ be a set of design measures such that $\xi(\Delta X) = \omega(\Delta X)$ for any $\Delta X \subset \operatorname{supp} \xi$ and $\xi(\Delta X) = 0$ otherwise.

A function $\psi(x, \xi)$ is said to separate sets X_1 and X_2 with respect to the measure $\omega(dx)$ if for any two sets $\Delta X_1 \subset X_1$ and $\Delta X_2 \subset X_2$ with equal nonzero measures (e.g., in the simplest case, area)

$$\int_{\Delta X_1} \psi(x, \xi) \omega(dx) \leq \int_{\Delta X_2} \psi(x, \xi) \omega(dx). \tag{4.3.2}$$

Theorem 4.3.1 *If assumptions (B1)-(B4) and (D) hold, then*

1. $\xi^* \in \Xi$ exists;

2. *a necessary and sufficient condition of* (Ψ, ω)-*optimality for* $\xi^* \in \overline{\Xi}$ *is that* $\psi(x, \xi^*)$ *separates* $X^* = \operatorname{supp} \xi^*$ *and* $X \setminus X^*$.

Proof: The results of the theorem are strongly related to the moment spaces theory, and the proof is based on corresponding ideas.

1. The existence of an optimal design follows from the compactness of the set of information matrices. The fact that at least one optimal design belongs to $\overline{\Xi}$ is a corollary of Liapunov's Theorem on the range of vector measures [cf. Karlin and Studden (1966), Chpt. VIII.12]. This theorem states in terms of information matrices that for any design measure $\xi(dx) \leq \omega(dx)$ and for the corresponding information matrix

$$M(\xi) = \int_X f(x) f^T(x) \, \xi(dx) = \int_X f(x) f^T(x) \varphi(x) \, \omega(dx),$$

where $\int_X \xi(dx) = \int_X \varphi(x) \omega(dx) = 1$ and $0 \leq \varphi(x) \leq 1$, there exists a subset $X' \subset X$ such that

$$\int_{X'} f(x) f^T(x) \, \omega(dx) = M(\xi)$$

and $\int_{X'} \omega(dx) = 1$. A measure $\xi'(dx)$ that fulfills $\xi'(dx) = \omega(dx)$ if $x \in X'$ and $\xi'(dx) \equiv 0$ otherwise can be considered as a design from $\overline{\Xi}$.

2. The necessity follows from the fact that if there exist subsets $\Delta X_1 \subset X^*$ and $\Delta X_2 \subset X \setminus X^*$ with nonzero measures such that

$$\int_{\Delta X_1} \psi(x, \xi^*) \, \omega(dx) > \int_{\Delta X_2} \psi(x, \xi^*) \, \omega(dx),$$

then deletion of ΔX_1 from the support set and subsequent inclusion of ΔX_2 causes a decrease of Ψ [cf. (2.3.16)]. This contradicts the optimality of ξ^*.

To prove the sufficiency we assume that $\xi^* \in \overline{\Xi}$ is non-optimal and $\xi \in \overline{\Xi}$ is optimal, i.e.,

$$\Psi[M(\xi^*)] > \Psi[M(\xi)] + \delta \tag{4.3.3}$$

for some $\delta > 0$. Let $\overline{\xi} = (1 - \alpha)\xi^* + \alpha\xi$; then the convexity of Ψ implies that

$$\begin{aligned} \Psi[M(\overline{\xi})] &\leq (1 - \alpha)\Psi[M(\xi^*)] + \alpha\Psi[M(\xi)] \\ &\leq (1 - \alpha)\Psi[M(\xi^*)] + \alpha\{\Psi[M(\xi^*)] - \delta\} \\ &= \Psi[M(\xi^*)] - \alpha\delta. \end{aligned} \tag{4.3.4}$$

On the other hand, assumption (B4) states that

$$\Psi[M(\overline{\xi})] = \Psi[M(\xi^*)] + \alpha \int_X \psi(x, \xi^*) \, \xi(dx) + o(\alpha).$$

Let

$$\operatorname{supp} \xi = (X^* \setminus D) \cup E$$

for $D \subset X^*$, $E \subset (X \setminus X^*)$, and $E \cap D = \emptyset$, where $\int_E \omega(dx) = \int_D \omega(dx)$ to assure that $\int_X \xi(dx) = \int_X \xi^*(dx) = 1$. Then we find that

$$\int_X \psi(x, \xi^*) \, \xi(dx) \tag{4.3.5}$$

$$= \int_{X^*} \psi(x, \xi^*) \, \omega(dx) + \int_E \psi(x, \xi^*) \, \omega(dx) - \int_D \psi(x, \xi^*) \, \omega(dx).$$

From assumption (B4) and in particular from (2.3.12), we find for $\overline{\xi} = \xi$ and $\alpha \to 0$ that

$$\int_X \psi(x, \xi)\,\xi(dx) = 0\,.$$

As a consequence,

$$\int_{X_*} \psi(x, \xi^*)\,\omega(dx) = \int_X \psi(x, \xi^*)\,\xi^*(dx) = 0\,.$$

From the assumption of separation follows that

$$\int_E \psi(x, \xi^*)\,\omega(dx) \geq \int_D \psi(x, \xi^*)\,\omega(dx)\,, \tag{4.3.6}$$

and we find that

$$\int_X \psi(x, \xi^*)\,\xi(dx) \geq 0\,.$$

This implies

$$\Psi[M(\overline{\xi})] \geq \Psi[M(\xi^*)] + o(\alpha)\,. \tag{4.3.7}$$

Comparison of (4.3.3) and (4.3.7) indicates a contradiction, and this completes the proof.

To summarize, let us note that in spite of its seemingly abstract form, *Theorem 4.3.1* contains the following, rather simple message. For the *D*-criterion, for instance, it tells us that the normalized variance of the predicted response $d(x, \xi^*)$ must at all support points of an optimal design ξ^* be greater than anywhere else. Of course, the same statement is true for any sensitivity function $\phi(x, \xi^*)$ from *Table 2.1*. Similar to the standard case, we have to allocate observations at points where we know least about the response. Moreover, if we are lucky enough to select a design which has this property, then it is an optimal design. Of course, we have to remember that $\xi^*(dx) = \omega(dx)$ at all support points.

In practical applications, $\xi^*(dx)$ can be used as follows. Optimal designs with directly constrained design measures occupy some subregions of X. One possibility is to allocate in an area ΔX a number

$$N^*(\Delta X) = \left[N \int_{\Delta X} \xi^*(dx) \right]^+$$

of sensors. They may be sited at the nodes of some uniform grid.

Iterative Procedure. Similar to the standard case, *Theorem 4.3.1* allows to develop a simple iterative procedure for constructing optimal designs. This theorem tells us that $\xi^*(dx)$ should be different from 0 in areas where $\psi(x, \xi^*)$ admits smaller values. Therefore, relocating some measure from areas with higher values to those with smaller values will hopefully improve ξ. This simple idea can be converted to a numerical procedure in many ways. We consider one that is very similar in its computer realization to an iterative procedure that is based on a first order algorithm.

(a) For a design $\xi_s \in \bar{\Xi}$, let $X_{1s} = \operatorname{supp} \xi_s$ and $X_{2s} = X \setminus X_{1s}$; find

$$x_{1s} = \arg \max_{x \in X_{1s}} \psi(x, \xi_s), \quad x_{2s} = \arg \min_{x \in X_{2s}} \psi(x, \xi_s), \qquad (4.3.8)$$

and two sets $D_s \subset X_{1s}$ and $E_s \subset X_{2s}$ such that $x_{1s} \in D_s$, $x_{2s} \in E_s$, and

$$\int_{D_s} \omega(dx) = \int_{E_s} \omega(dx) = \alpha_s$$

for some $\alpha_s > 0$.

(b) Construct ξ_{s+1} such that

$$\operatorname{supp} \xi_{s+1} = X_{1,s+1} = (X_{1s} \setminus D_s) \cup E_s \,.$$

If α_s is element of a sequence that fulfills

$$\lim_{s \to \infty} \alpha_s = 0, \quad \sum_{s=1}^{\infty} \alpha_s = \infty,$$

and if the conditions of *Theorem 4.3.1* and assumption (B4′) from Section 3.2 hold, then $\{\Psi(\xi_s)\}$ converges to ξ^* as defined by (4.3.1). The proof of this statement is almost identical to the proof of *Theorem 3.2.1*.

In most problems, $\omega(dx) = \varphi(x)dx$. These cases may be converted to the case where $\varphi(x) \equiv c$ for some real c by an appropriate transformation of the coordinates. For $\varphi(x) \equiv c$, all integrals may be replaced by sums over some regular grid elements in the computerized version of the iterative procedure. If these elements are fixed and D_s and E_s coincide with the grid elements, then the above iterative procedure may be considered as a special version of the exchange algorithm with one simple constraint: No grid element can have more than one support point and the weights of all support points are the same, i.e., N^{-1}. Thus, there is no need to create a special software; minor modifications of widely used packages can make the job.

Example: Polynomial Regression. The D-optimal design for fitting $\eta(x, \theta) = \theta_1 + \theta_2 x + \theta_3 x^2$ is to be estimated. Let the design measure $\xi(dx)$ be uniform on $[-1, 1]$ and restricted by the condition $\int \omega(dx) = \int \omega_0 dx = 2$. The restriction implies that $\omega_0 = 1$. Obviously, the design ξ has to be symmetric, and observations in the surrounding of $x = -1$, 0, and 1 are essential. Thus, the structure of X^* can be written as the union $[-1, -a] \cup [-a', a'] \cup [a, 1]$ of intervals for suitably chosen numbers a and a'. From $\int \xi(dx) = \int \omega_0 dx = 1$ follows $a' = a - 0.5$; a must be element of $(0.5, 1]$. The maximum of the determinant $|M|$ of the information matrix is reached at $a^* \doteq 0.71$.

The sensitivity function $d(x, \xi^*)$ of the corresponding design ξ^* (see Figure 4.1) illustrates *Theorem 4.3.1*: The variance of the response function $d(x, \xi^*)$ is larger for all x's in X^* than for any x in $[-1, 1] \setminus X^*$; the two sets are separated by $d(x, \xi^*)$ and consequently by $\psi(x, \xi^*)$, as expected due to part 2 of the theorem.

Figure 4.1: D-optimal design for a polynomial of order 2 on $[-1, 1]$, restricted by $\int \omega \, dx = 2$. Sensitivity function $d(x, \xi^*)$ over x. The bold line represents X^*.

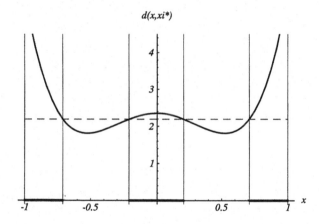

Chapter 5

Special Cases and Applications

5.1 Designs for Time-Dependent Models

In this section we assume that the independent variables $x_1, \ldots, x_k$ (and the response variable y) are observed in the space of variables t that are out of the experimenter's control but are known. Typically, t stands for time; but t may correspond to temperature, atmospheric pressure, wind direction in environmental studies; to age, living conditions, gender in biological studies, etc. We associate t with time because time-dependent models are of great practical relevance. But no specific features of "time" are crucial for the following. The reader may adapt the results to other situations.

The experimenter controls (and is able to vary) the variables x at any time t. Series of observations $x(t_i)$ that are taken at fixed and known time points $t_1, \ldots, t_q$ may be correlated. We use the term "trajectory" to emphasize the time-dependence of $x(t)$. Such trajectories are the elements of our design space. They will be characterized as $(k \times q)$-matrices

$$\boldsymbol{x}_i^T = (x_i(t_1), \ldots, x_i(t_q)).$$

The corresponding basis functions of the standard model form an $(m \times q)$-matrix

$$\boldsymbol{F}(\boldsymbol{x}_i) = (f[x_i(t_1), t_1], \ldots, f[x_i(t_q), t_q]),$$

and the response is a q-vector

$$\boldsymbol{y}_i^T = (y_i(t_1), \ldots, y_i(t_q)),.$$

Note that the situation which we discuss in this section differs essentially from that treated in Section 2.6 where we also have assumed that independent variables are out of the experimenter's control. In contrast to here, these variables (denoted in Section 2.6 by u) are explicit arguments of the basis functions and are unknown to the experimenter.

The Model and Analytical Results. By use of the q-vectors

$$
\begin{aligned}
\boldsymbol{y}_i^T &= (y_i(t_1), \ldots, y_i(t_q)), \\
\boldsymbol{\varepsilon}_i^T &= (\varepsilon_i[x_i(t_1), t_1], \ldots, \varepsilon_i[x_i(t_q), t_q]),
\end{aligned}
$$

the $(k \times q)$-matrix

$$\boldsymbol{x}_i^T = (x_i(t_1), \ldots, x_i(t_q)),$$

and the $(m \times q)$-matrix

$$\boldsymbol{F}(\boldsymbol{x}_i) = (f[x_i(t_1), t_1], \ldots, f[x_i(t_q), t_q]),$$

we define the model as

$$\boldsymbol{y}_i = \boldsymbol{F}^T(\boldsymbol{x}_i)\,\theta + \boldsymbol{\varepsilon}_i \qquad (5.1.1)$$

for $i = 1, \ldots, n$, where

$$\mathrm{E}\{\boldsymbol{\varepsilon}_i\} = 0, \quad \mathrm{E}\{\boldsymbol{\varepsilon}_i\boldsymbol{\varepsilon}_j^T\} = \delta_{ij}C(\boldsymbol{x}_i). \qquad (5.1.2)$$

As (5.1.2) states, correlation between observations on the same trajectory may occur but observations from different trajectories are not correlated. The latter fact is crucial for this approach because it assures the additivity of the information matrices for observations from different trajectories [cf. (1.1.16)]. We avoid at this point to generalize our model to the case of continuous time in order to keep the discussion at a simple level.

To estimate θ, we have to observe the response along appropriately chosen trajectories $\boldsymbol{x}_i$. The information matrix for model (5.1.1) is given by

$$M = \sum_{i=1}^{n} p_i M(\boldsymbol{x}_i), \qquad (5.1.3)$$

where $p_i = r_i/N$, r_i is the number of trajectories $\boldsymbol{x}_i$ $(N = \sum_{i=1}^{n} r_i)$, and

$$qM(\boldsymbol{x}_i) = \boldsymbol{F}(\boldsymbol{x}_i)C^{-1}(\boldsymbol{x}_i)\boldsymbol{F}(\boldsymbol{x}_i). \qquad (5.1.4)$$

The normalizing multiplier q is convenient if $C(\boldsymbol{x}_i) \equiv I$. In this case, it allows to link the results of this section to those from the standard theory. It also helps to avoid infinite growth of $M(\boldsymbol{x}_i)$ when $q \to \infty$. However, in cases where the errors are correlated, e.g., for long memory processes, this normalization can lead to a vanishing $M(\boldsymbol{x}_i)$ if $q \to \infty$. Therefore, some precaution is necessary if q becomes large.

In terms of the continuous design theory,

$$M(\xi) = \int_{\boldsymbol{X}} M(\boldsymbol{x})\,\xi(d\boldsymbol{x}), \qquad (5.1.5)$$

where $\xi(d\boldsymbol{x})$ is a probability (or design) measure with support points $\boldsymbol{x} \in \boldsymbol{X}$. Let us note that the statement $\boldsymbol{x} \in \boldsymbol{X}$ means that $x(t) \in X(t)$ for $t = t_1, \ldots, t_q$. Typically, the design set $\boldsymbol{X}$ is defined by

$$\boldsymbol{x}^T \boldsymbol{x} \le 1 \quad \text{or} \quad \max_t |x(t)| \le 1, \qquad (5.1.6)$$

where each component $x(t)$ is one-dimensional. Obvious extensions can be made for multi-dimensional cases. For the more general case of continuous time, the scalar product in (5.1.6) must be replaced by the corresponding integral.

In previous chapters we have never used any special features of x or X except the assumption that X is a compact set. Therefore, whenever we can show that X is compact, all results of Chapters 2 and 3 may be applied. However, for applying these results we have to make some obvious changes. For instance, expressions of the form $f^T(x)Af(x) = \operatorname{tr} f(x)f^T(x)A$ must be replaced by $\operatorname{tr} M(x)A$. This change is intuitively quite natural if we recall that in the standard case $m(x) = f(x)f^T(x)$ is the information matrix of an observation made at x. As an example for the transfer of results from the standard case we state that for the D-criterion, according to *Theorem 2.4.1* (see also *Theorem 2.3.2*), a necessary and sufficient condition for ξ^* to be optimal is the inequality

$$\operatorname{tr} M(x)D(\xi^*) \le m\,. \tag{5.1.7}$$

Example: First Order Autocorrelation. Let

$$y_i = \theta^T f(x_i) + \varepsilon_i\,, \tag{5.1.8}$$

where $|x_{ij}| \le 1$, and the vector of errors has components $\varepsilon_{i1} = \nu_{i1}$ and $\varepsilon_{ij} = \rho\varepsilon_{i,j-1} + \nu_{ij}$ for $1 < j \le q$, with $\mathrm{E}\{\nu_i\} = 0$ and $\mathrm{E}\{\nu_i\nu_i^T\} = I$. In matrix notation, we have

$$L\varepsilon_i = \nu_i\,,$$

where

$$L = \begin{pmatrix} 1 & 0 & 0 & \dots & 0 & 0 \\ -\rho & 1 & 0 & \dots & 0 & 0 \\ \vdots & \vdots & \vdots & \ddots & \vdots & \vdots \\ 0 & 0 & 0 & \dots & -\rho & 1 \end{pmatrix}.$$

We find that

$$\mathrm{E}\{\varepsilon_i\varepsilon_i^T\} = C = (L^T L)^{-1}$$

and, consequently,

$$C^{-1} = L^T L\,.$$

A D-optimal design ξ^* has to fulfill the inequality (5.1.7) Therefore, we have to analyze the behavior of the function $\operatorname{tr} M(x)D(\xi)$. From (5.1.4) follows that

$$\begin{aligned} q\operatorname{tr} M(x)D(\xi) &= \operatorname{tr} F(x)LL^T F^T(x)D(\xi) \\ &= \operatorname{tr} L^T F^T(x)D(\xi)F(x)L \\ &= \sum_{j=1}^{q} d(x_j, \xi) + \rho^2 \sum_{j=2}^{q-1} d(x_j, \xi) - 2\rho \sum_{j=1}^{q-1} d(x_j, x_{j+1}, \xi)\,, \end{aligned} \tag{5.1.9}$$

where $d(x_j, x_k, \xi) = f^T[x(t_j), t_j]D(\xi)f[x(t_k), t_k]$ and $d(x_j, \xi) = d(x_j, x_j, \xi)$. To apply (5.1.7), consider the behavior of function (5.1.9) for the simple linear regression with

$$f^T(x, t) = (1, x)\,.$$

In this case (5.1.9) is a second order polynomial with respect to $x_1, \dots, x_q$. Therefore, all components of a support vector of the optimal design ξ^* must be ± 1. Since the design region is symmetric with respect to zero, the information matrix of an optimal design must be diagonal. Hence

$$d(x_{j-1}, x_j, \xi^*) = D_{11}(\xi^*) + D_{22}(\xi^*)x_{j-1}x_j \,.$$

If $\rho > 0$, the term $d(x_{j-1}, x_j, \xi^*)$ contributes a positive amount only if $x_{j-1} = -x_j$. Therefore, for positive ρ (to avoid complications we assume that j is even) an optimal design can be found which consists of only one trajectory ($p = 1$) with

$$\pmb{x}^* = \{x_j = (-1)^j\}_1^q \,.$$

This is not the only optimal design. For instance, the design that gives equal weights to the two trajectories $\pmb{x}^*$ and $-\pmb{x}^*$ is also optimal.

If $\rho < 0$ it can be shown that the design with two support trajectories $x_{1j} \equiv 1$ and $x_{2j} \equiv -1$ and with equal weights $p_1 = p_2 = 0.5$ is a unique optimal design.

Parameterization of Trajectories. If q is large, we may try to reduce the dimension of the optimization problem by introducing a parametric approximation of the trajectories $x(t)$. Let us, for example, assume that $x(t)$ can be approximated by

$$x(t) = \varphi(u, t)\,, \tag{5.1.10}$$

where the vector u contains κ parameters. Clearly, $\varphi(u, t)$ has to be sufficiently flexible to approximate $x(t)$. We assume that $x(t)$ is scalar in order to avoid technicalities. The information matrix can be represented by

$$M(\xi) = \int_U F(u)C^{-1}(u)F(u)\,\xi(du)\,, \tag{5.1.11}$$

with $U = \{u : \varphi(u, t) \in X(t), t \in T\}$, and a necessary and sufficient condition for D-optimality is

$$\operatorname{tr} M(u)M^{-1}(\xi^*) \leq m\,. \tag{5.1.12}$$

Note that the dimension of this optimization problem is κ, the dimension of u.

In addition to reducing the dimension of the optimization problem, the approximation (5.1.10) allows us to take constraints on $x(t)$ into account. For instance, suppose that the costs of changing the level of a control variable are proportional to the distance between $x(t_i)$ and $x(t_{i+1})$. Then we may add to (5.1.12) the constraint (we assume that $t_{i+1} - t_i \to 0$ for the sake of simplicity)

$$\int_0^T \int_{X(t)} \left(\frac{dx}{dt}\right)^2 \xi(dx|t)\, dt \leq L_1\,,$$

i.e., we limit the "average squared slope". If we look for "shorter" trajectories, the constraint

$$\int_0^T \int_{X(t)} \sqrt{1 + x^2(t)}\, \xi(dx|t)\, dt \leq L_2$$

might be appropriate; this constraint means that we limit the average length of the trajectories. In these formulas, $\xi(dx|t)$ stands for the "profile" of $\xi(dx)$ at time t. The trajectories are observed in the interval $[0, T]$.

Generalized Trajectories as Conditional Distributions. Assume that in every series of observations the errors are independent and have equal variances, i.e., $C(\boldsymbol{x}_i) = I_q$. Then

$$M = \sum_{i=1}^{n} p_i M(\boldsymbol{x}_i) = \sum_{i=1}^{n} \sum_{j=1}^{q} p_i q^{-1} f[x_i(t_j), t_j] f^T[x_i(t_j), t_j]. \tag{5.1.13}$$

In terms of continuous time we get

$$M(\xi) = \int_T \int_{X(t)} M[x(t)]\, \xi(dx, dt). \tag{5.1.14}$$

If we define

$$\xi(dt) = \int_{X(t)} \xi(dx, dt),$$

the identity $\xi(dt) \equiv c$ corresponds to the uniform distribution of the times of observation, i.e., $t_{i+1} - t_i$ is the same for all i. In other words, $\xi(dt)$ defines the "density" of times of observation.

Thus, we can restate the optimization problem (2.3.6) in the form

$$\xi^* = \arg \min_{\xi} \Psi[M(\xi)] \quad s.t. \quad \xi(dt) = \xi^0(dt), \tag{5.1.15}$$

where $\xi^0(dt)$ is assumed to be given. Introduction of

$$\xi^0(dt) = \sum_{j=1}^{q} q^{-1} \delta(t - t_j)\, dt,$$

where $\delta(t)$ is the standard δ-function, brings us back to (5.1.13).

The time-dependent measure $\xi^*(dx|t)$ can be considered as the optimal generalized trajectory. It is closely related to the generalized optimal functions ("controls") as considered in optimal control theory.

The additivity of M [cf. (5.1.13)] is essential for the approach that is considered in this section. Noting that

$$M(\boldsymbol{x}) = \int_T M[x(t), t] \xi(dt), \tag{5.1.16}$$

we can replace (5.1.7) by the following necessary and sufficient condition for optimality of $\xi^*(dx|t)$:

$$d(x, t, \xi^*) \leq d(x^*, t, \xi^*) \tag{5.1.17}$$

for any $x^* \in \operatorname{supp} \xi^*(dx|t)$, where

$$d(x, t, \xi) = f^T(x, t) D(\xi) f(x, t) = \operatorname{tr} M(x, t) M^{-1}(\xi).$$

Applying (5.1.17), we can find that the optimal design does not depend on t for some special cases. For instance, let

(a) $\eta(x,t,\theta) = \mu + \gamma^T f_1(x) + \beta^T f_2(t)$, where $\theta^T = (\mu, \gamma^T, \beta^T)$ or

(b) $\eta(x,t,\theta) = \theta^T f(x,t)$, where $f(x,t) = f_1(x) \otimes f_2(t)$,

and let $X(t) \equiv X$ in both cases. Then, the optimal design consists of a generalized trajectory which is generated by the measure $\xi^*(dx|t)$ and which coincides with the support points of an optimal design for the following models, respectively:

(a) $\eta(x,\theta_1) = \mu + \gamma^T f_1(x)$, $\theta_1^T = (\mu, \gamma^T)$, $x \in X$, and

(b) $\eta(x,\theta_1) = \theta_1^T f_1(x)$, $x \in X$.

Numerical Procedures. In general, the iterative procedures as described in Chapter 3 may be applied without changes. For the D-criterion, we have to find

$$\boldsymbol{x}_s = \arg\max_{\boldsymbol{x}} \operatorname{tr} M(\boldsymbol{x})D(\xi_s) \qquad (5.1.18)$$

for the "forward" step. No special problems are caused by (5.1.18) as compared with the standard case. However, for practical values of q and multi-dimensional x, the computational expenses may be rather high. Parameterization [cf. (5.1.10)] usually helps to reduce the computational effort. The dimension of the optimization problem coincides with the number of parameters of the approximation.

Another intuitively attractive simplification of the computations is possible when the necessary and sufficient condition (5.1.7) can be replaced by (5.1.17). In this case (5.1.18) splits into q separate optimization problems:

$$x_s(t_j) = \arg\max_{x \in X(t_j)} d(x, t_j, \xi_s) \qquad (5.1.19)$$

and $x_s(t_2)$ has to be the local maximum of $d(x, t_2, \xi_s)$ nearest to $x_s(t_1)$, $x_s(t_3)$ has to be the local maximum of $d(x, t_3, \xi_s)$ nearest to $x_s(t_2)$ and so on.

Example: Polynomial Trajectory. The response function is a second order polynomial regression

$$\eta(x,t,\theta) = \theta_1 + \theta_2 x + \theta_3 t + \theta_4 xt + \theta_5 x^2 + \theta_6 t^2$$

with $-1 \leq x(t) \leq 1$ and $T = \{t_i = -1 + 0.1(i-1), i = 1, \ldots, 21\}$. For deriving a D-optimal design, we use simplified version of the iterative procedure based on (5.1.17): The step length α is the same for all steps and points with small weight are deleted. Figures 5.1 and 5.2 show the variance of the response function $d(x,t,\xi)$ for the initial design ξ_0 and for the D-optimal design ξ^*, respectively.

The initial design ξ_0 consists of two trajectories. For both, $x_{it} = 0$ for $t = \pm 0.9$, ± 0.6, ± 0.3 and 0; $x_{1t} = 1$ and $x_{2t} = -1$ otherwise. Thus, one third of the observations are taken at each of the points $x = -1$, 0, and 1. The corresponding marginal measure $\xi(dx)$ coincides with the D-optimal design for the univariate second order polynomial model. The rounded version of the optimal design ξ^* consists also of two trajectories; for both, observations are taken in $x_{it} = 0$ at times $t = \pm 0.8$, ± 0.2, and ± 0.1; $x_{1t} = 1$ and $x_{2t} = -1$ otherwise. Going from the initial to the optimal design, the number

Figure 5.1: Variance of the response function $d(x, t, \xi)$ for the initial design ξ_0.

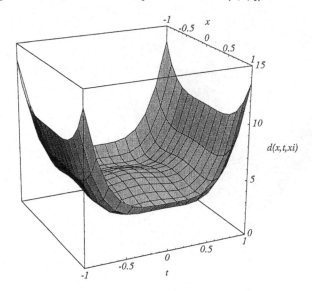

of observations at $x = -1$ and 1 increases for small and large t; correspondingly, the number of observations at $x = 0$ decreases. For times close to 0, the observations become more concentrated at $x = 0$. In other words, the trajectories move towards the local maximum of $d(x, t, \xi_0)$. The determinant $|D|$ decreases from 698.8 to 625.7. The inspection of the figures allows to cautiously state that ξ^* satisfies (5.1.17). The comparison of figures 5.1 and 5.2 shows that the initial design perfoms quite well as compared to the computed design and may be recommended for practical use. However, it is far from sure that such a recommendation can be given in general.

5.2   Regression Models with Random Parameters

We consider experiments where observations y_{ij} are taken from the j-th object, $j = 1, \ldots, J$, under conditions x_i, $i = 1, \ldots, n$. To give examples, we associate the index j with individuals in medico-biological studies, with a date in meteorology, or with a tested item in engineering.

Let us assume that the response function of the standard model form an has the same form for all objects, i.e.,

$$y_{ij} = f^T(x_i)\theta_j + \varepsilon_{ij}, \tag{5.2.1}$$

where the vector $\theta_j = (\theta_{1j}, \ldots, \theta_{mj})^T$ of (individual) parameters reflects the specific features of the j-th object. Let the variability of θ_j be described by a probability distribution with mean θ_0, the vector of global or population parameters, and dispersion

Figure 5.2: Variance of the response function $d(x, t, \xi)$ for the optimal design ξ^*.

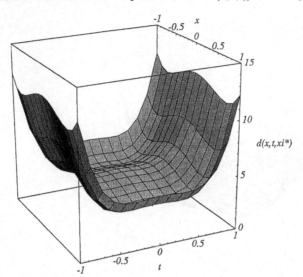

matrix D_0. We further assume that the observation errors are uncorrelated, have zero mean and variance σ^2, and that they are independent of θ_j.

The Estimation Problem. Model (5.2.1) is usually called a regression model with random coefficients or a regression model of the second kind. The aim of the statistical analysis of such a model can be

(a) the estimation of the individual parameters θ_j,

(b) the estimation of the global parameters $\theta_0 = E\{\theta\}$,

(c) the estimation of σ^2 and D_0

or mixtures of these. Of course, the actual aim (and design) depends on our *a priori* knowledge of the model. In general, for different j we may use different designs for the data collection. For simplicity, we confine ourselves to the case where identical designs are used for all j.

Table 5.1 shows estimators for all three situations of prior knowledge. In this table

$$\underline{M}_j \;\equiv\; \underline{M} = \sigma^{-2} N M = \sigma^{-2} N \sum_{i=1}^{n} p_i f(x_i) f^T(x_i),$$

$$\underline{Y}_j \;=\; \sigma^{-2} N Y_j = \sigma^{-2} \sum_{i=1}^{n} r_i \overline{y}_{ij} f(x_i),$$

	θ_0, σ^2, and D_0 known	σ^2 and D_0 known θ_0 unknown	θ_0^2, σ^2, and D_0 unknown
$\hat\theta_j$	$(\underline{D}_0^{-1} + \underline{M})^{-1}(\underline{D}_0^{-1}\theta_0 + \underline{Y}_j)$	$\underline{M}^{-1}\underline{Y}_j$	$M^{-1}Y_j$
D_{θ_j}	$\underline{D}_0^{-1} + \underline{M}$	$\underline{M}^{-1}$	$\hat\sigma^2 N^{-1}M^{-1}$
$\hat\theta_0$	—	$J^{-1}\underline{M}^{-1}\sum_{j=1}^J \underline{Y}_j$	$J^{-1}N^{-1}M^{-1}\sum_{i=1}^n Y_j$
$D_{\hat\theta_0}$	—	$J^{-1}(\underline{L}_0 + \underline{M}^{-1})$	$J^{-1}(\underline{D}_0 + \hat\sigma^2 N^{-1}M^{-1})$
$\hat\sigma^2$	—	—	$J^{-1}(N-m)^{-1}\sum_{j=1}^J \sum_{i=1}^n$ $\sum_{\ell=1}^{r_i}\left(y_{ijt} - \hat\theta_j f(x_i)\right)^2$
$\underline{D}_0$	—	—	$(J-1)^{-1}\sum_{j=1}^J$ $(\hat\theta_j - \hat\theta_0)(\hat\theta_j - \hat\theta_0)^T - \hat\sigma^2 N^{-1}M^{-1}$

Table 5.1: Estimators and dispersion matrices for three levels of *a priori* knowledge.

$$\bar{y}_{ij} = r_i^{-1} \sum_{\ell=1}^{r_i} y_{ij\ell}.$$

In these formulas, the number r_i of repeated observations is the same for $j = 1, \ldots, J$, and $p_i = r_i/N$, $N = \sum_{i=1}^{n} r_i$. Note, that $E_\theta E_\varepsilon \{(\hat{\theta}_j - \theta_j)\} = 0$ and

$$D_{\theta_j} = E_\theta E_\varepsilon \{(\hat{\theta}_j - \theta_j)(\hat{\theta}_j - \theta_j)^T\}, \qquad (5.2.2)$$

where the subscripts θ and ε indicate that the expectation corresponds to θ and ε, respectively. For the global parameters, $E_\theta E_\varepsilon \{\hat{\theta}_0\} = \theta_0$ and

$$D_{\hat{\theta}_0} = E_\theta E_\varepsilon \{(\hat{\theta}_0 - \theta_0)(\hat{\theta}_0 - \theta)^T\} \qquad (5.2.3)$$

The estimators $\hat{\theta}_j$ and $\hat{\theta}_0$ that are shown in *Table 5.1* are best linear unbiased estimators (b.l.u.e.) in the sense of (5.2.2) and (5.2.3), respectively.

Optimality Criteria for two Estimation Problems. From *Table 5.1* is clear that the optimization problem

$$\xi^* = \arg\min_\xi \Psi[D_0^{-1} + M(\xi)], \qquad (5.2.4)$$

where $D_0 = \sigma^{-2} N \underline{D}_0$, has to be considered if we want to estimate the individual parameters θ_j. Note that this is identical to the Bayesian design problem [cf. (2.6.5)]. If the global parameter θ_0 is of interest, then

$$\xi^* = \arg\min_\xi \Psi[(D_0 + M^{-1}(\xi))^{-1}]. \qquad (5.2.5)$$

Similar to the Bayesian design problem, (5.2.4) often leads to optimal designs with singular information matrix. Intuitively, this is obvious because θ_0 and D_0 may give us a fair knowledge about θ_j. If in case of (5.2.5) all components of θ_0 are of interest, the optimal design must be regular.

Technically, both (5.2.4) and (5.2.5) are special cases of (2.3.6), and all results of Chapters 2 and 3 may be applied. For instance, if

$$\Psi\left[(D_0 + M^{-1})^{-1}\right] = \log|D_0 + M^{-1}|, \qquad (5.2.6)$$

a necessary and sufficient condition for ξ^* to be optimal for the estimation of θ_0 is (cf. *Theorem 2.3.2*) that

$$f^T(x)M^{-1}(\xi^*)[D_0 + M^{-1}(\xi^*)]^{-1}M^{-1}(\xi^*)f(x)$$
$$\leq \ \text{tr}[D_0 + M^{-1}(\xi^*)]^{-1}M^{-1}(\xi^*). \qquad (5.2.7)$$

As the criterion is similar to the D-criterion, the design ξ^* is often called D-optimal. By means of

$$(A + B)^{-1} = A^{-1} - A^{-1}(A^{-1} + B^{-1})^{-1}B^{-1}, \qquad (5.2.8)$$

[cf. result (C) of the Appendix A] we may rewrite (5.2.7) as

$$d(x, \xi^*) - d_{tot}(x, \xi^*) \leq m - \text{tr}[D_0^{-1} + M(\xi^*)]^{-1}M(\xi^*), \qquad (5.2.9)$$

where $d(x, \xi) = f^T(x)M^{-1}(\xi)f(x)$ and $d_{tot}(x, \xi) = f^T(x)[D_0^{-1} + M(\xi)]^{-1}f(x)$.

The condition (5.2.9) is quite different from what we get for a D-optimal design for estimating θ_j [cf. (5.2.4)]:

$$\Psi[D_0^{-1} + M(\xi)] = -\log|D_0^{-1} + M(\xi)|. \tag{5.2.10}$$

A necessary and sufficient condition for that is

$$d_{tot}(x, \xi^*) \leq \mathrm{tr}[D_0^{-1} + M(\xi^*)]^{-1}M(\xi^*). \tag{5.2.11}$$

Thus, in the latter case, the observations must be allocated at points where the variance of the estimated response is largest, while for estimating the global parameters the allocation of observations is based on the difference between two variances: $d(x, \xi)$ is the variance when no *a priori* information is involved, whereas $d_{tot}(x, \xi)$ uses our knowledge of D_0. Obviously, $d(x, \xi) \geq d_{tot}(x, \xi)$ for any x and ξ.

5.3 Mixed Models and Correlated Observations

In environmental monitoring, the response may be determined by the location of the sensor but also by the weather conditions; these, however, are random and similar in a wide region. An appropriate model must take random effects and correlated observations into account.

Let the model be defined as

$$y_{ij} = \theta^T f(x_i) + u_j(x_i) + \varepsilon_{ij}, \tag{5.3.1}$$

where $i = 1, \ldots, n$ and $j = 1, \ldots, J$. Similar to Section 5.2, the observations y_{ij} are taken under conditions x_i from the j-th object that may be a date in environmental monitoring or in meteorology, an individual in medico-biological studies, or a tested item in engineering. We skip the index j when this does not cause any confusion.

The model (5.3.1) contains two different "generators" of randomness. The "observational" error is represented by ε. The variable u describes the deviation of the observed response from $\theta^T f(x)$ due to causes that may depend on x. Let x indicate, e.g., the coordinates of some sensor or observing station in environmental monitoring. Then, u may describe the deviations that are caused by, e.g., weather fluctuations on the scale of the whole region X.

We admit that observations at different sites may be correlated. We assume that this dependence is described through the covariance kernel for the random variables $u_i = u(x_i)$ $(\mathrm{E}\{u(x_i)\} = 0)$:

$$\mathrm{E}\{u(x_i)u(x_j)\} = K(x_i, x_j) \tag{5.3.2}$$

Observational errors are represented by ε. We assume that these errors are uncorrelated for different sites x_i and for different observation times. They have mean zero and the same variance σ^2. More complicated situations may be transformed to this case [cf. (1.3.1) and (1.3.2)].

Let the kernel $K(x_i, x_j)$ be defined on $Z \otimes Z$, where Z is compact and $x_i, x_j \in X \subset Z \subset R^k$. We assume that $K(x_i, x_j)$ is associated with the Hilbert space L_2, i.e., for any $\varphi \in L_2$ exists the integral $\int_Z K(x_i, x_j)\varphi(x_j)dx_j$. Then [see, e.g., Kanwal (1971), Chpt. 7], the kernel can be represented as a uniformly and absolutely convergent sum:

$$K(x_i, x_j) = \sum_{l=1}^{\infty} \lambda_l \varphi_l(x_i)\varphi_l(x_j) \qquad (5.3.3)$$

and the series $\sum_{l=1}^{\infty} \lambda_l$ is convergent.

Obviously λ_l must diminish not slower than l^{-1}. In many cases the decay is much faster. This fact allows to hope that for practical needs we can use the approximation

$$K(x_i, x_j) \simeq K_p(x_i, x_j) = \sum_{l=1}^{p} \lambda_l \varphi_l(x_i)\varphi_l(x_j) \qquad (5.3.4)$$

for some moderate p.

We introduce random parameters $\gamma^T = (\gamma_1, \ldots, \gamma_p)$ such that $E\{\gamma\} = 0$ and

$$E\{\gamma\gamma^T\} = \Lambda \qquad (5.3.5)$$

with $\Lambda_{\alpha\beta} = \lambda_\alpha \delta_{\alpha\beta}$, for $\alpha, \beta = 1, \ldots, p$. If the approximation (5.3.4) is acceptable, (5.3.1) can be substituted by the model

$$y_i = \theta^T f(x_i) + \gamma^T \varphi(x_i) + e_i. \qquad (5.3.6)$$

Optimal Prediction without Trend. We want to predict the response y for points from some given set X_{pr}. For that purpose, we have to select sites from X so that these predictions are most effective, i.e., are as accurate as possible. Two aspects contribute to the uncertainty of the prediction: Uncertainty of the estimators for θ, and uncertainty that is caused by the randomness of u and ε. We discuss first the case where $\theta^T f(x) \equiv 0$, i.e., the model contains no trend. In this case, the prediction is entirely determined by the covariances between the components of u at various locations.

Let the design ξ_n consist of support points $(x_1, \ldots, x_n)$ and equal weights $p_i = n^{-1}$. We introduce the $(n \times n)$-matrix $K(\xi_n)$ with elements

$$\{K(\xi_n)\}_{ij} = K(x_i, x_j), \qquad (5.3.7)$$

the n-vectors $K(x, \xi_n)$ with i-th component $K(x, x_i)$, and $y^T = (y_1, \ldots, y_n)$. It is easy to check that $E\{y\} = 0$ and

$$E\{yy^T\} = V(\xi_n) = \sigma^2 I + K(\xi_n). \qquad (5.3.8)$$

It is known [see Ripley (1981), Chpt. 4.4] that

$$\hat{y}(x) = V^T(x, \xi_n)V^{-1}(\xi_n)y \qquad (5.3.9)$$

is the best linear unbiased predictor for $y(x)$; here, $V(x, \xi_n) = K(x, \xi_n)$ if $x \neq x_i$, and $V(x, \xi_n) = \sigma^2 + K(x, \xi_n)$ if $x = x_i$. If x does not coincide with any support point of ξ_n, then the variance of this predictor is equal to

$$\begin{aligned} \text{Var}\{y(x) - \hat{y}(x)\} &= \text{E}\{(y(x) - \hat{y}(x))^2\} \qquad\qquad (5.3.10) \\ &= V(x, x) - V^T(x, \xi_n) V^{-1}(\xi_n) V(x, \xi_n) \\ &= \sigma^2 + K(x, x) - K^T(x, \xi_n)[\sigma^2 I + K(\xi_n)]^{-1} K(x, \xi_n). \end{aligned}$$

Otherwise it obviously diminishes. The expectation is taken with respect to u and ε.

The most commonly used objective functions related to (5.3.10) are the maximal variance of prediction

$$Q_1(\xi_n) = \max_{x \in X_{pr}} \text{Var}\{y(x) - \hat{y}(x)\} \qquad\qquad (5.3.11)$$

and the average variance of prediction

$$Q_2(\xi_n) = \int_{X_{pr}} \text{Var}\{y(x) - \hat{y}(x)\} dx. \qquad\qquad (5.3.12)$$

Consequently, the design problem may be stated as

$$\xi_n^* = \arg\min_{\xi_n} Q(\xi_n), \qquad\qquad (5.3.13)$$

where Q stands either for Q_1 or Q_2. A weight function may be added. However, the changes in the final results will be obvious and we therefore prefer to analyze the simpler case.

Direct solution of (5.3.13) is beyond ideas developed in this text. However, the approximate version of (5.3.13) may be attacked by means of methods that were considered above, in particular in Sections 2.6 and 5.2.

We want to emphasize at this point that it is crucial for the rest of this section that $\sigma^2 \neq 0$, and that the approximation (5.3.4) and subsequently (5.3.6) are valid. If $\theta = 0$, i.e., the model contains no trend, then (5.3.6) is reduced to

$$y_i = \gamma^T \varphi(x_i) + e_i. \qquad\qquad (5.3.14)$$

Within this approximation

$$\text{E}\{yy^T\} = V(\xi_n) = \sigma^2 I + K(\xi_n) = \sigma^2 I + \Phi^T(\xi_n) \Lambda \Phi(\xi_n), \qquad\qquad (5.3.15)$$

where $\Phi(\xi_n) = (\varphi(x_1), \ldots, \varphi(x_n))$. The best linear predictor for γ is (cf. *Table 2.1*)

$$\hat{\gamma} = \sigma^{-2}[\underline{M}(\xi_n) + \Lambda^{-1}]^{-1} \Phi(\xi_n) y, \qquad\qquad (5.3.16)$$

where $\underline{M}(\xi_n) = \sigma^{-2} \Phi(\xi_n) \Phi^T(\xi_n)$. The dispersion matrix of the difference $\hat{\gamma} - \gamma$ is

$$D(\xi_n) = \text{E}\{(\hat{\gamma} - \gamma)(\hat{\gamma} - \gamma)^T\} = [\underline{M}(\xi) + \Lambda^{-1}]^{-1}; \qquad\qquad (5.3.17)$$

expectation is taken with respect to u and ε.

Let us choose

$$\tilde{y}(x) = \varphi^T(x)\hat{\gamma}$$

as a predictor for $y(x)$. Actually, both $\tilde{y}(x) = \varphi^T(x)\hat{\gamma}$ and $\hat{y}(x)$ from (5.3.9) can be used as a predictor for $u(x)$, and all the results may be applied if $u(x)$ is of prime interest. On an intuitive level, it is obvious that $\hat{y}(x)$ and $\tilde{y}(x)$ must coincide to an extent that is determined by the approximation (5.3.4). Indeed, using the identity (C) from the Appendix A, we can check that if x is not element of supp ξ_n, then

$$\begin{aligned}
\tilde{y}(x) &= \sigma^{-2}\varphi^T(x)[\underline{M}(\xi_n) + \Lambda^{-1}]^{-1}\Phi(\xi_n)y \\
&= \varphi^T(x)\Lambda\Phi(\xi_n)[\Phi^T(\xi_n)\Lambda\Phi(\xi_n) + \sigma^2 I]^{-1}y \\
&= K(x,\xi_n)[\sigma^2 I + K(\xi_n)]^{-1}y = \hat{y}(x)\,, \quad\quad\quad (5.3.18)
\end{aligned}$$

and

$$\begin{aligned}
\operatorname{Var}\{\tilde{y}(x) - y(x)\} &= \sigma^2 + \varphi^T(x)[\Lambda^{-1} + \underline{M}(\xi_n)]^{-1}\varphi(x) \\
&= \sigma^2 + \varphi^T(x)\Lambda\varphi(x) \\
&= \varphi^T(x)\Lambda\Phi(\xi_n)[\sigma^2 I + \Phi^T(\xi_n)\Lambda\Phi^T(\xi_n)]^{-1}\Phi^T\xi_n)\Lambda\varphi(x) \\
&= \sigma^2 + K(x,x) - K(x,\xi_n)[\sigma^2 I + K(\xi_n)]^{-1}K(x,\xi_n) \\
&= \operatorname{Var}\{\hat{y}(x) - y(x)\}\,. \quad\quad\quad (5.3.19)
\end{aligned}$$

The latter identity allows to replace (5.3.13) by the more familiar optimization problem

$$\xi_n^* = \arg\min_{\xi_n} \Psi[D(\xi_n)]\,, \quad\quad\quad (5.3.20)$$

where $D(\xi_n)$ is defined in (5.3.17). If we want to use the maximal variance of prediction [cf. (5.3.11)], we obtain

$$\Psi(D) = \max_{x \in X_{pr}} \varphi^T(x)D\varphi(x)\,. \qu\quad\quad (5.3.21)$$

For the average variance of prediction [cf. (5.3.12)], we get

$$\Psi(D) = \operatorname{tr} AD\,, \qu\quad\quad (5.3.22)$$

where

$$A = \int_{X_{pr}} \varphi(x)\varphi^T(x)dx\,.$$

Although (5.3.20) looks like a typical design problem, we have to recall that at every point x_i the design ξ_n has just one observation, i.e., $p_i \equiv N^{-1}$ is not subject of minimization.

Let us admit in the following that r_i observations are taken for each j. This means that we have to replace $\sigma^2 I$ by $\sigma^2 J$ in all formulas which have been derived in this section [see, e.g., (5.3.18) and (5.3.19)]; J has elements $J_{ij} = r_i^{-1}\delta_{ij}$.

In a monitoring network setting, the time that is spent to observe some characteristics or the area where contamination was measured may be used instead of the number r_i of observations. Thus, the weights can quite naturally be considered as continuous. As a consequence, the introduction of the continuous design

$$\xi = \{p_i, x_i\}_1^n$$

with $x_k \in X$ and $0 \le p_i \le 1$ arises as a quite natural, and crucial, step.

Necessary and Sufficient Conditions. These conditions for the optimality of designs are stated in terms of the covariance function. Let us redefine the matrix $D(\xi)$ as

$$D(\xi) = [\sigma^{-2} N M(\xi) + \Lambda^{-1}]^{-1}, \tag{5.3.23}$$

where

$$M(\xi) = \int_X \varphi(x) \varphi^T(x) \xi(dx),$$

and consider a continuous version of (5.3.20)

$$\xi^* = \min_\xi \Psi[D(\xi)], \tag{5.3.24}$$

which is a special case of the optimization problems analyzed in the previous section.

Let us continue the analysis of the design problem for the criteria $Q_1(\xi)$ and $Q_2(\xi)$, i.e., the criteria based on the maximal and average variance of prediction, respectively. From *Theorem 2.3.2* follows that a design ξ^* is optimal for criterion (5.3.22), i.e., minimizes the average variance of prediction, if and only if for all $x \in X$

$$\varphi^T(x) D(\xi^*) A D(\xi^*) \varphi(x) \le \operatorname{tr} M(\xi^*) D(\xi^*) A D(\xi^*), \tag{5.3.25}$$

and the equality is true for all support points of ξ^*.

For criterion (5.3.21) we introduce the additional assumption that $X_{pr} = X$. Then a design ξ^* is minimax if and only if for all $x \in X$

$$\varphi^T(x) D(\xi) \varphi(x) \le \operatorname{tr} M(\xi^*) D(\xi^*), \tag{5.3.26}$$

and the equality holds at all support points of ξ^*. Moreover, the design problem (5.3.20) is equivalent to

$$\xi^* = \arg \min_\xi |D(\xi)|. \tag{5.3.27}$$

These results are rather obvious outcomes of the convex design theory. However, we may come to more interesting conclusions if these results are reformulated in terms of the covariance function

$$C(x, x'|\xi) = K(x, x') - K(x, \xi)[\sigma^2 J(\xi) + K(\xi)]^{-1} K(x', \xi). \tag{5.3.28}$$

This function is the covariance of the predictions $\tilde{y}(x)$ and $\tilde{y}(x')$. Note also that the identity (B2) of the Appendix A allows us to write

$$D(\xi) = \Lambda - \Lambda \Phi(\xi)[\sigma^{-2} J(\xi) + \Phi(\xi)^T \Lambda \Phi(\xi)]^{-1} \Phi(\xi)^T \Lambda. \tag{5.3.29}$$

Combining (5.3.15), (5.3.25), (5.3.26), and (5.3.29) with (5.3.28) leads us to conditions for the optimality of designs which are, for convenience, formulated as a theorem.

Theorem 5.3.1 *The design ξ^* minimizes the average variance of prediction if and only if for all $x \in X$*

$$\int_{X_{pr}} C^2(x, x'|\xi)\, dx' \le \int_X \int_{X_{pr}} C^2(x, x'|\xi^*)\, \xi^*(dx)\, dx' \qquad (5.3.30)$$

and the equality holds for all support points of ξ^*.

Theorem 5.3.2

1. *The design ξ^* is minimax, if and only if for all its support points x_i^**

$$\mathrm{Var}\{y(x_i^*) - \tilde{y}(x_i^*)|\xi^*\} = \max_{x \in X} \mathrm{Var}\{y(x) - \tilde{y}(x)|\xi^*\}. \qquad (5.3.31)$$

2. *Minimax designs coincide with D-optimal designs, i.e., $\xi^* = \arg\min_\xi |D(\xi)|$ or equivalently*

$$\xi^* = \arg\max_\xi |\sigma^2 J(\xi) + K(\xi)||\sigma^2 J(\xi)|^{-1}. \qquad (5.3.32)$$

To see (5.3.32) we make use of

$$|\sigma^{-2} N M(\xi) + \Lambda^{-1}| = \frac{|\sigma^2 J(\xi) + K(\xi)|}{|\Lambda||\sigma^2 J(\xi)|}.$$

Note that $\mathrm{Var}\{y(x) - \tilde{y}(x)|\xi\} = C(x, x|\xi)$. Intuitively, both results are very natural. In the first case the support points of ξ^* must be chosen from those where $\tilde{y}(x)$ is most correlated in average with $\tilde{y}(x_{pr})$ for $x_{pr} \in X_{pr}$.

In the minimax case, the support points must be chosen from points where the prediction is worst. In many cases, optimal weights are equal for all support points, i.e., $J(\xi^*) = (Np)^{-1} I$ and (5.3.32) is equivalent to the maximization of the determinant of matrix $\sigma^2 J(\xi) + K(\xi)$ which is the covariance matrix of "observations" at points $x_1^*, \ldots, x_n^*$. In other words, to find a minimax design ξ^* we have to select n points such that the determinant of their covariance matrix is the greatest possible.

In the last paragraph, "observation" stands for $\bar{y}_i = \frac{1}{r_i} \sum_{j=1}^{r_i} y_{ij}$. Let us remind that all above results are valid if

$$K(x_i, x_j) = \sum_{l=1}^p \lambda_l \varphi_l(x_i) \varphi_l(x_j)$$

for $p < \infty$. If our intention is to use the afore-mentioned theorems for nonsingular covariance kernels, then we have to investigate the asymptotic behavior of ξ^* for $p \to \infty$. Very little is known in this area.

Numerical Procedures. Obviously, all results from Chapter 3 may be applied to construct optimal designs for the regression problem (5.3.14). Most of these results may be reformulated in terms of the covariance function $C(x, x'|\xi)$.

For instance, for the minimax case we can use the following iterative procedure (cf. Section 3.3).

(a) Let the design ξ_s consist of N points, each with weight N^{-1}; find

$$x_s^+ = \arg\max_{x \in X} C(x, x|\xi)$$

and add this point to design ξ_s:

$$\xi_{Ns}^+ = \{\xi_{Ns} + x_s^+\}. \tag{5.3.33}$$

(b) Find

$$x_s^- = \arg\min_{x \in X_s^+} C(x, x|\xi_{Ns}^+),$$

where $X_s^+ = \operatorname{supp} \xi_{Ns}$; delete x_s^- from ξ_{Ns}^+, i.e., construct

$$\xi_{N,s+1} = \{\xi_{Ns}^+ - x_s^-\}. \tag{5.3.34}$$

In (5.3.33) and (5.3.34), "+" and "−" stand for adding or deleting the corresponding support points.

Optimal Design in the Presence of Trend. Introduction of the trend $\theta^T f(x)$ makes all results technically more complex, but they are based on exactly the same ideas that were used above.

Let us assume that the trend parameters θ are of prime interest. Then some function of the dispersion matrix of their estimators is to be minimized. Let

$$M_{11}(\xi) = \int f(x) f^T(x) \, \xi(dx),$$

$$M_{12}(\xi) = \int f(x) \varphi^T(x) = M_{21}^T(\xi),$$

$$M_{22}(\xi) = \int \varphi(x) \varphi^T(x) \, \xi(dx),$$

$$M_{022} = \sigma^2 N^{-1} \Lambda^{-1},$$

and

$$\Delta(\xi) = \begin{pmatrix} \Delta_{11}(\xi) & \Delta_{12}(\xi) \\ \Delta_{21}(\xi) & \Delta_{22}(\xi) \end{pmatrix} = \begin{pmatrix} M_{11}(\xi) & M_{12}(\xi) \\ M_{21}(\xi) & M_{22}(\xi) + M_{022} \end{pmatrix}^{-1}.$$

The matrix $\Delta_{11}(\xi)$ is the standardized dispersion matrix of $\hat{\theta}$; cf. (1.1.26) for

$$D_0 = \begin{pmatrix} 0 & 0 \\ 0 & \Lambda \end{pmatrix}.$$

Therefore,

$$\xi^* = \arg\min_{\xi} \Psi[\Delta_{11}(\xi)].$$

For instance, for the D-criterion we have to minimize $\log|\Delta_{11}(\xi)|$. From *Theorem 2.3.2*, follows immediately that a necessary and sufficient condition for ξ^* to be optimal is that

$$f^T(x)\Delta(\xi^*)f(x) - f_2^T(x)[M_{22}(\xi^*) + M_{022}]^{-1}f_2(x)$$
$$\leq \operatorname{tr}\Delta(\xi^*)M(\xi^*) - \operatorname{tr}[M_{22}(\xi^*) + M_{022}]^{-1}M_{22}(\xi^*). \tag{5.3.35}$$

Note that $f^T(x)\Delta(\xi)f(x)$ is the normalized variance of $\hat{\theta}f(x)$, and $f_2^T(x)[M_{22}(\xi) + M_{022}]^{-1}f_2(x)$ may be considered as a normalized variance of the best linear estimator for the regression model with the same observational errors but with only the response function $\gamma^T\varphi(x)$.

Condition (5.3.35) may be presented in a form similar to (5.3.31). A curious reader might engage in the not complicated but rather long and tedious algebra to derive this result.

5.4 Design for "Contaminated" Models

Obviously, any model is only an approximation to the real relationship between the involved variables of the data-generating process. If the design region X is small enough, a simple linear model with $f^T(x) = (1, x_1, \ldots, x_k)$ will probably be appropriate. Often the region X is intentionally shrunk to assure the applicability of the selected model. Actually in most applications the choice of X is a compromise between the practitioner's faith that a model still "works" in X and the physical ability to vary controlled variables. Of course, we have to take the possibility of discrepancies between the specified model and the data-generating process into account. If this is the case, the model often is denoted as "contaminated".

In this section we try to develop some tools that may help to understand how X must be chosen, or, vice versa, how complex the response function can be. The model validity is a crucial point.

The Model. Let us introduce a (contamination) term $\gamma^T\varphi(x)$ in the standard model:

$$y = \theta^T f(x) + \underline{\gamma}^T\varphi(x) + \varepsilon; \qquad (5.4.1)$$

$\underline{\gamma}^T\varphi(x)$ describes the deviation of the standard model from the true one. The structure of this new model is the same as that of the old one and still simple. Nevertheless, we can use this model for analyzing phenomena which may occur when the discrepancy between the contaminated model and the true relation is significant.

We use the standard least squares method for estimating the parameters θ of the contaminated model. Then we can investigate the effect of ignoring the (contamination) term $\underline{\gamma}^T\varphi(x)$. It is well-known that $\hat{\theta}$ is biased. The normalized mean squared error matrix of the least squares estimator $\hat{\theta}$ is

$$
\begin{aligned}
R(\xi) &= \mathrm{E}\{(\hat{\theta}-\theta)(\hat{\theta}-\theta)^T\} \\
&= M_{11}^{-1}(\xi) + M_{11}^{-1}(\xi)M_{12}(\xi)BM_{21}(\xi)M_{11}^{-1}(\xi) \\
&\equiv D(\xi) + W(\xi), \qquad\qquad (5.4.2)
\end{aligned}
$$

where

$$M_{11}(\xi) = \int f(x)f^T(x)\,\xi(dx),$$

$$M_{12}(\xi) = \int f(x)\varphi^T(x) = M_{21}^T(\xi),$$

$$M_{22}(\xi) = \int \varphi(x)\varphi^T(x)\,\xi(dx),$$

and $B = \gamma\gamma^T$ with $\gamma = \sqrt{N}\gamma/\sigma$. In $R(\xi) = D(\xi) + W(\xi)$, $D(\xi)$ is the normalized dispersion matrix of $\hat{\theta}$, whereas $W(\xi)$ is the corresponding normalized bias matrix.

To assess the effect of $\gamma \neq 0$, let us assume that γ is given. Obviously, we can transform (5.4.1) into the standard model by substituting $y' = y - \gamma^T\varphi(x)$ for y. Let us consider the analogue of the D-criterion and define an optimal design as

$$\xi^* = \arg\min_\xi \log|R(\xi)|. \qquad (5.4.3)$$

It is sometimes called D_R-optimal when it is necessary to distinguish it from other optimal designs.

At the first glance, (5.4.3) is another case where the results of Chapter 2 may be applied. However, one important assumption is not fulfilled: $\log|R(\xi)|$ is not a convex function of the information matrix $M_{11}(\xi)$ [cf. (B1) in Section 2.3]. Moreover, it is in general not a convex function of the design ξ. Note that (5.4.3) defines the standard D-optimal design if the standard model without contamination term is adequate.

The Model Validity Range. Given the design region X, we may define the "model validity range" as the set (range) Γ_0 of all γ, such that

$$\min_\xi \log|R(\xi)| \leq \min_\xi \log|\overline{D}(\xi)|, \qquad (5.4.4)$$

where $\overline{D}(\xi)$ is the normalized dispersion matrix of the least squares estimator for the vector θ in the frame of the extended model (5.4.1), i.e.,

$$[\overline{D}(\xi)]^{-1} = \overline{M}(\xi) = M_{11}(\xi) - M_{12}(\xi)M_{22}^{-1}(\xi)M_{12}(\xi)$$

(see Section 2.1). We assume that of M_{11} and M_{22} are regular. As long as the actual γ is in this range, we can gain, in terms of (5.4.4), by using the standard model without contamination term.

Obviously, for orthogonal D_R-optimal designs ($M_{12} = 0$), the set Γ_0 becomes infinite. That is the reason why orthogonality was considered as one of the most important features in many classical works on experimental design. Unfortunately, orthogonal designs may be constructed only in very special cases: E.g., X has to be symmetric (cube, sphere), and $f(x)$ and $\varphi(x)$ must possess some symmetry as well. Orthogonality can be achieved through a proper selection of design if, e.g., $f^T(x) = (1, x_1, \ldots, x_k)$, $\varphi^T(x) = (x_1 x_2, \ldots, x_{k-1} x_k)$, and X is either a cube or a sphere. However, if $\varphi(x)$ is extended so that it includes also quadratic terms $x_1^2, \ldots, x_k^2$, then orthogonal designs cannot be constructed, and we have to find new ways to decide about the range of the parameters γ for which the standard model without contamination term is appropriate.

Another concept for the model validity range is as follows: Given γ, the model validity range is the set X_0 such that for any $X' \subset X_0$

$$\min_{\xi \in \Xi(X')} \log |R(\xi)| \leq \min_{\xi \in \Xi(X')} \log |\overline{D}(\xi)|, \tag{5.4.5}$$

where $\Xi(X')$ is the set of all designs defined on X'. The definitions (5.4.4) and (5.4.5) are denoted as Γ-validity range and X-validity range, respectively; the notions are dual. Of course, both concepts are strongly determined by the optimality criterion on which they are based. We have to discuss the model validity range in the sense of the D-, the linear, or any other optimality criterion.

Model validity ranges may equally be based on the minimax, or on the Bayesian version of (5.4.5), replacing $\log |R(\xi)|$ by $\max_{\gamma \in \Gamma_0} \log |R(\xi)|$ or $\int_{\Gamma_0} \log |R|(\xi)| \pi_0(\gamma) d\gamma$, respectively, where Γ_0 and $\pi_0(\gamma)$ are assumed to be given.

To get a feeling for model validity ranges the reader might exercise with the model $\theta_1 + \theta_2 x + \gamma x^2$, $|x| < 1$. Straightforward algebra leads to

$$|R(\xi)| = [1 + \gamma^2 m_2^2(\xi)] m_2(\xi),$$

where $m_2(\xi) = \int x^2 \xi(dx)$. Any design that is symmetrical with respect to the origin and has the second moment $m_2^* = \min\{1, |\gamma|^{-1}\}$ is D_R-optimal. Examples are

$$\xi_1^* = \left\{ \begin{array}{cc} -\sqrt{m_2^*} & \sqrt{m_2^*} \\ \frac{1}{2} & \frac{1}{2} \end{array} \right\}, \ \xi_2^* = \left\{ \begin{array}{ccc} -1 & 0 & 1 \\ \frac{1}{2}m_2^* & 1 - m_2^* & \frac{1}{2}m_2^* \end{array} \right\}. \tag{5.4.6}$$

Applying (5.4.4) and using *Theorem 2.3.2* and *Table 2.1* we find $\min_\xi \log |\overline{D}(\xi)| = 4$ and

$$\xi_2^* = \left\{ \begin{array}{ccc} -1 & 0 & 1 \\ \frac{1}{4} & \frac{1}{2} & \frac{1}{4} \end{array} \right\}.$$

The Γ-validity range turns out to be $|\gamma| \leq 2$.

Optimal Designs for Contaminated Models. The straightforward approaches for constructing optimal designs work only in very simple cases. The above-mentioned lack of convexity prevents us from using results of Chapters 2 and 3. But some of them are still valid. We proceed with the D_R-criterion, leaving to the reader to extend the results for other criteria.

By calculating the directional derivatives, we can conclude that a necessary condition for ξ^* to be D_R-optimal is the inequality

$$\max_{x \in X}[d_1(x, \xi^*) + d_2(x, \xi^*)] \leq m - \frac{k(\xi^*)}{k(\xi^*) + 1}, \tag{5.4.7}$$

where

$$\begin{array}{rcl} d_1(x, \xi) & = & f^T(x) M_{11}^{-1}(\xi) f(x), \\ d_2(x, \xi) & = & \dfrac{\zeta(x, \xi) [\zeta(x, \xi) - 2r(x)]}{k(\xi) + 1}, \end{array}$$

$$\zeta(x,\xi) = \int d_1(x,x',\xi)r(x')\xi(dx'),$$
$$d_1(x,x',\xi) = f^T(x)M_{11}^{-1}(\xi)f(x'),$$
$$r(x) = \gamma^T\varphi(x),$$

and

$$k(\xi) = \int\int r(x)d_1(x,x',\xi)r(x')\xi(dx)\xi(x') \qquad (5.4.8)$$
$$= \gamma^T M_{21}(\xi)M_{11}^{-1}(\xi)M_{12}(\xi)\gamma.$$

At each support point of the optimal design, the function $d_1(x,\xi^*)+d_2(x,\xi^*)$ achieves its upper bound.

Note that (5.4.7) is only a necessary condition. This is due to the lack of convexity of $\log|R(\xi)|$ with respect to ξ. In addition to that, (5.4.7) includes functions that depend on the unknown γ. The consequence is that we can learn from (5.4.7) something about the structure of optimal designs, but we cannot construct one which can immediately be applied in practice. The dependence of the optimal design on γ can be handled by adopting, e.g., the Bayesian approach. All results and discussions can be generalized to the case where $B = \gamma\gamma^T$ is replaced everywhere by $B = \int \gamma\gamma^T \pi_0(\gamma)d\gamma = E\{\gamma\gamma^T\}$; here, $\pi_0(\gamma)$ is the *a priori* distribution of γ.

More can be done if the optimization problem (5.4.3) is replaced by an approximate version. Let us note that

$$\log|R(\xi)| = S_1(\xi) + S_2(\xi), \qquad (5.4.9)$$

where $S_1(\xi) = \log|M_{11}^{-1}(\xi)|$ and $S_2(\xi) = \log\left[1 + \gamma^T M_{21}(\xi)M_{11}^{-1}(\xi)M_{12}(\xi)\gamma\right]$. This partitioning is an evident corollary of identity (B1) of the Appendix A. If we prefer to think in Bayesian terms, $\gamma^T A\gamma = \mathrm{tr}\,\gamma\gamma^T A$ must be replaced by $\mathrm{tr}\,BA$.

Observing that

$$\log(1+u) = u - \frac{u^2}{2(1+w)^2} \qquad (5.4.10)$$

for some $0 \leq w \leq u$, we may conclude that

$$\log|R(\xi)| \leq \log|M_{11}^{-1}(\xi)| + k(\xi) \equiv Q(\xi),$$

where $k(\xi)$ is defined in (5.4.8). Thus,

$$\min_{\xi}\log|M_{11}^{-1}(\xi)| \leq \min_{\xi}\log|R(\xi)| \leq \min_{\xi}Q(\xi). \qquad (5.4.11)$$

If the designs

$$\xi_L = \arg\min_{\xi}\log|M_{11}^{-1}(\xi)|,$$

$$\xi_U = \arg\min_{\xi}Q(\xi)$$

are known, (5.4.11) may be used to evaluate the "goodness" of any design ξ in the sense of the D_R-criterion. In addition, we may use another corollary of (5.4.10):

$$\frac{k^2(\xi)}{2(1 + k(\xi))^2} \leq Q(\xi) - \log |R(\xi)| \leq \frac{k^2(\xi)}{2} . \tag{5.4.12}$$

It is known that

$$[(1 - \alpha)A_1 + \alpha A_2] [(1 - \alpha)B_1 + \alpha B_2]^{-1} [(1 - \alpha)A_1 + \alpha A_2]^T$$

$$\leq (1 - \alpha)A_1 B_1^{-1} A_1^T + \alpha A_2 B_2^{-1} A_2^T , \tag{5.4.13}$$

where A_j is an arbitrary $(r \times s)$-matrix and B_j is a positive-definite $(s \times s)$-matrix ($j = 1, 2$) [see (D) in the Appendix A]. This inequality shows that the functions $k(\xi)$ and, in consequence, $Q(\xi)$ are convex. This, together with *Theorem 2.3.2*, immediately leads us to a necessary and sufficient condition: A design ξ_U minimizes $Q(\xi)$ if and only if

$$d_1(x, \xi_U) + d_0(x, \xi_U) \leq m - k(\xi_U), \tag{5.4.14}$$

where $d_0(x, \xi) = \phi(x, \xi)(\phi(x, \xi) - 2r(x))$; $d_1(x, \xi)$ and $r(x)$ are defined in the context of (5.4.7).

Formulas (5.4.11) and (5.4.14) allow to use the techniques from Chapter 3 without major changes. Of course, some of the designs will depend on the unknown parameters γ. Therefore, it is recommended to construct optimal designs under various assumptions about γ that are compatible with intuition and *a priori* knowledge.

A more elaborated approach can be developed if we apply results that are discussed in Section 2.6 for optimality criteria which depend on uncontrolled variables, and in particular *Theorem 2.6.1*.

Constrained Optimization. Formulas (5.4.11) and (5.4.12) are essentially based on properties of the objective function $\log |R(\xi)|$. They hold for the D_R-criterion but are not generally valid for other criteria. An alternative approach that makes use of the idea of constrained optimization helps to overcome, at least partly, these difficulties.

So far, our objective was to find

$$\xi^* = \arg \min_\xi \Psi[R(\xi)] , \tag{5.4.15}$$

and the D_R-criterion was used as an illustration. From (5.4.15) and the definition of $R(\xi) = D(\xi) + W(\xi)$ [cf. (5.4.2)], it is obvious that if $D(\xi)$ is minimized, it is usually to the expense of $W(\xi)$. This is also true vice versa. Suppose that we want to minimize the estimation bias, and at the same time, the random part of uncertainty in estimation has to be kept under control. Then a logical step is to look for designs that minimize some function of $W(\xi)$, but ensure that $M^{-1}(\xi)$ will not be too large. Thus, an optimal design is defined as

$$\xi_{wd}^* = \arg \min_\xi \Psi[W(\xi)] \quad \text{s.t.} \quad \Phi[D(\xi)] \leq A , \tag{5.4.16}$$

where Φ is an appropriate function and A is a suitably chosen constant. Ideally, $\Psi[W(\xi)]$ and $\Phi[D(\xi)]$ are chosen to be convex functions of ξ in order to allow the formulation of corresponding necessary and sufficient conditions and the construction of procedures.

The dual problem might also be of interest. An optimal design under a mean squared error based criterion with a constraint on the bias term is defined as

$$\xi_{dw}^{*} = \arg\min_{\xi} \Phi[D(\xi)] \quad \text{s.t.} \quad \Psi[W(\xi)] \leq B, \tag{5.4.17}$$

where B again is a suitably chosen constant.

Problems (5.4.16) and (5.4.17) look very similar to the constrained optimization problem (4.2.1) where an optimal design is to be found subject to a general constraint. However, even if $\Psi(W)$ is a convex function of W, the function $\Psi[W(\xi)]$ may be a non-convex function of ξ.

The situation is simple if we use the D-criterion. Then (5.4.9) leads us to

$$\Phi(D) = \log|D| = \log|M_{11}^{-1}| \tag{5.4.18}$$

and

$$\Psi(W) = \gamma^{T} M_{21} M_{11}^{-1} M_{12} \gamma. \tag{5.4.19}$$

In the Bayesian case we use $\Psi(W) = \operatorname{tr} B M_{21} M_{11}^{-1} M_{12}$ instead of (5.4.19).

The combination of (5.4.18) and (5.4.19) with either (5.4.16) or (5.4.17) immediately leads us to a special case of the general constrained optimization problem (4.2.1). Therefore, all results of Section 4.2 can be applied and no new developments are needed. For instance, the extended version of *Theorem 4.1.1* provides us with a necessary and sufficient condition of optimality of ξ_{dw}^{*} for the optimization problem (5.4.17), viz., the existence of $u^{*} > 0$ such that for all $x \in X$

$$d_{1}(x, \xi_{dw}^{*}) + u^{*} d_{2}(x, \xi_{dw}^{*}) \leq m - u^{*} B. \tag{5.4.20}$$

In the case of (5.4.16), the inequality

$$
\begin{aligned}
u^{*} d_{1}(x, \xi_{wd}^{*}) &+ d_{2}(x, \xi_{wd}^{*}) \\
&\leq u^{*} m - \gamma^{T} M_{21}(\xi_{wd}^{*}) M_{11}^{-1}(\xi_{wd}^{*}) M_{12}(\xi_{wd}^{*}) \gamma
\end{aligned} \tag{5.4.21}
$$

must be fulfilled instead of (5.4.20). It is interesting to note that in the latter case an optimal design does not depend on the length of the vector γ but only on its direction.

For the D-criterion is it essential that $\log|R(\xi)|$ can be split into two terms [cf. (5.4.9)]. What can we do in other cases? We discuss only the linear criterion. Other criteria may be treated similarly.

To mimic (5.4.9), we choose

$$\Psi(D) = \operatorname{tr} AD$$

and
$$\Phi(W) = \operatorname{tr} AW\,.$$

Minimization of $\Phi(W)$ is not easier than the minimization of $\operatorname{tr} AR$: The function $\Phi[W(\xi)]$ is not convex with respect to ξ and we therefore cannot apply the theory from Section 4.2. We can improve the situation by changing the structure of $\Psi(D)$ and $\Phi(W)$. For instance, instead of $\operatorname{tr} AW$ we can consider the relative quadratic losses:

$$\Phi(W) = \operatorname{tr} M_{11} AW\,. \tag{5.4.22}$$

To make the discussion simpler, we choose $A = I$. Then the optimization problem (5.4.17) becomes

$$\xi_{dw}^* = \arg\min_{\xi} \operatorname{tr} M_{11}^{-1}(\xi) \quad \text{s.t.} \quad \gamma^T M_{21}(\xi) M_{11}^{-1}(\xi) M_{12}(\xi)\gamma \le B\,. \tag{5.4.23}$$

Obviously, (5.4.23) satisfies the conditions of the extended version of *Theorem 4.1.1* and we may conclude that a necessary and sufficient condition for optimality of ξ_{dw}^* is the existence of $u^* > 0$ such that for all $x \in X$

$$f^T(x) M_{11}^{-2}(\xi_{dw}^*) f(x) + u_2^* d_2(x, \xi_{dw}^*) \le \operatorname{tr} M_{11}^{-1}(\xi_{dw}^*) - u^* B\,. \tag{5.4.24}$$

Thus, if we use (5.4.22) instead of $\operatorname{tr} AW$ for $\Phi(W)$, all results from Section 4.2 may be applied. We suggest the reader to repeat this exercise for the dual counterpart of (5.4.23).

All numerical procedures that are discussed in Section 4.2 may be applied to the optimization problems of this section in a very straightforward way.

5.5 Model Discrimination

In practical situations, knowledge of the functional form $\eta(x)$ is often not certain as it was assumed so far in this text. Instead, we might be confident that one of several models will be adequate but we do not know which one of them. This makes clear that designs are needed which give us high power in discriminating among models, i.e., in deciding which of the models in question is the most adequate one.

Note that model discrimination, i.e., decision which of several models is the adequate one, is related to deciding whether a model is contaminated (cf. Section 5.4). However, there is a crucial difference between the two concepts. In the latter case, we have doubts and want to check whether a suspected contamination term should be included in the model. Model discrimination treats several models equally and without preference for any of the potentially adequate specifications.

In the following, we will assume that all models under consideration depend on the same set of independent variables; the methods will be discussed for the case that we have to discriminate between just two models.

Let the two models have the structure

$$y = \eta(x) + \varepsilon\,; \tag{5.5.1}$$

we discuss a generalized version of the standard model (1.1.1). The response function $\eta(x)$ is one of two known functions $\eta_1(x, \theta_1)$ or $\eta_2(x, \theta_2)$, where θ_1 and θ_2 are elements from $\Omega_1 \subset R^{m_1}$ and $\Omega_2 \subset R^{m_2}$, respectively. An experiment is to be performed on order to decide which one of the two models is adequate. Optimal designs for such experiments are our subject.

The results of this section will be based on least squares estimators. For most of these results, the linearity of the models is not essential. To define least squares estimators for the parameters θ_j, $j = 1, 2$, we introduce the sum of squares [cf. the notation of Section 1.1]

$$v_j(\theta_j) = \sum_{i=1}^{n} p_i[\bar{y}_i - \eta_j(x, \theta_j)]^2$$

with $p_i = r_i/N$. Then, the least squares estimators of θ_j is defined as

$$\hat{\theta}_j = \arg\min_{\theta_j \in \Omega_j} v_j(\theta_j). \tag{5.5.2}$$

Statistical methods of model discrimination are generally based on the comparison of the sum of squares $v_1(\hat{\theta}_1)$ and $v_2(\hat{\theta}_2)$. Without loss of generality, we consider only differences

$$d_{21} = v_2(\hat{\theta}_2) - v_1(\hat{\theta}). \tag{5.5.3}$$

A Basic Optimality Criterion. The optimal design for discriminating between the models will depend upon which model is the true one. Let us suppose that $\eta(x) = \eta_1(x, \theta_1)$ is the true response function. Obviously, we are the more certain that η_1 represents the true model, the larger the difference $d_{21} = v_2(\hat{\theta}_2) - v_1(\hat{\theta})$ is. The distribution of d_{21} is mainly determined by the sum of squared deviations

$$\Delta_{21}(\xi) = \sum_{i=1}^{n} p_i[\eta(x_i) - \eta_2(x_i, \hat{\theta}_2)]^2, \tag{5.5.4}$$

where

$$\hat{\theta}_2 = \arg\min_{\theta_2 \in \Omega_2} \sum_{i=1}^{n} p_i[\eta(x_i) - \eta_2(x_i, \theta_2)]^2; \tag{5.5.5}$$

$\Delta_{21}(\xi)$ is a measure for the lack of fit when $\eta_2(x_i, \hat{\theta}_2)$ is substituted for $\eta(x)$. If Ω_2 coincides with R^{m_2}, $\eta_2(x, \theta_2) = \theta_2^T f_2(x)$, and the errors are independently and normally distributed, then $\Delta_{21}(\xi)$ is proportional to the noncentrality parameter of the χ^2 distribution of $v_2(\hat{\theta}_2)$ [see Rao (1973), Chpt. 3b]. If $N \to \infty$, then the difference d_{21} converges almost surely to $\Delta_{21}(\xi)$ under mild assumptions; the independence of errors is the most important one.

If η_1 represents the true model the design

$$\xi^* = \arg\max_{\xi} \Delta_{21}(\xi) \tag{5.5.6}$$

that maximizes $\Delta_{21}(\xi)$ will give us the strongest hint to a lack of fit of the second model.

Locally Optimal Designs. The optimization problem (5.5.6) is a special case of the optimization problem (2.6.10) based on the minimax criterion with

$$\Psi(\xi, \theta) = -\int [\eta(x) - \eta_2(x, \theta)]^2 \xi(dx) ; \qquad (5.5.7)$$

here, the integral with respect to $\xi(dx)$ replaces the weighted sum in (5.5.4). Using the notation of this section, we reformulate (2.6.10) as

$$\xi^* = \arg \max_{\xi} \min_{\theta_2 \in \Omega_2} \Delta_{21}(\xi, \theta_2) , \qquad (5.5.8)$$

with

$$\Delta_{21}(\xi, \theta_2) = \int [\eta(x) - \eta_2(x, \theta_2)]^2 \xi(dx) . \qquad (5.5.9)$$

Obviously, *Theorem 2.6.1* can be applied to criterion (5.5.9) which is linear with respect to ξ. To assure that all other assumptions of this theorem hold, it is sufficient to assume that $\eta(x)$ and $\eta_2(x, \theta_2)$, for all $\theta_2 \in \Omega_2$, are continuous functions of $x \in X$. The analogue of the function $\psi(x, u, \xi)$ is

$$\Delta_{21}(\xi, \theta_2) - \varphi_{21}(x, \theta_2, \xi),$$

where $\varphi_{21}(x, \theta_2, \xi) = [\eta(x) - \eta_2(x, \theta_2)]^2$ is the sensitivity function. *Theorem 2.6.1* together with other results from Chapter 2 leads to the following theorem.

Theorem 5.5.1

  1. *A necessary and sufficient condition for a design ξ^* to be applied is the existence of a measure ζ^* such that*

$$\phi_{21}(x, \xi^*) \le \Delta_{21}(\xi^*) , \qquad (5.5.10)$$

where $\phi_{21}(x, \xi) = \int_{\Omega_2(\xi)} \phi_{21}(x, \theta_2, \xi) \zeta(d\theta_2)$, $\int_{\Omega_2(\xi)} \zeta(d\theta_2) = 1$, and

$$\Omega_2(\xi) = \{\theta_2 : \theta_2(\xi) = \arg \min_{\theta_2 \in \Omega_2} \Delta_{21}(\xi, \theta_s)\} . \qquad (5.5.11)$$

  2. *The function $\phi_{21}(x, \xi^*)$ achieves $\Delta_{21}(\xi^*)$ at all support points of ξ^*.*

  3. *The set of optimal designs is convex.*

Often, such optimal designs are called T-optimal where T stands for testing. If (5.5.11) has a unique solution then

$$\phi_{21}(x, \xi^*) = [\eta(x) - \eta_2(x, \theta_2^*)]^2$$

with $\theta_2^* = \theta_2(\xi^*)$; this makes (5.5.10) simple for understanding and verification. In particular, we notice that all observations have to be made at points where the function $\eta_2(x, \theta_2^*)$ deviates the most from $\eta_t(x)$.

Numerical Procedures. In applications, to get a feeling for the structure of a T-optimal design, we construct designs for a few values of $\theta_1 \in \Omega_1$ [if $\eta_1(x, \theta_1)$ is assumed to be true]. Analytical methods are possible only in special cases. In most situations, we use analogues of numerical procedure from Chapter 3. For instance, a simple first order algorithm may be based on the following iteration step:

(a) Given ξ_s, find

$$\theta_{2s} = \arg \min_{\theta_2 \in \Omega_2} \int [\eta(x) - \eta_2(x, \theta_2)]^2 \xi_s(dx),$$
$$x_s = \arg \max_{x \in X} [\eta(x) - \eta_2(x, \theta_{2s})]^2.$$

(b) Choose α_s with $0 \le \alpha_s \le 1$ and construct

$$\xi_{s+1} = (1 - \alpha_s)\xi_s + \alpha_s \xi(x_s).$$

The corresponding iterative procedure converges to an optimal design under mild conditions. For instance, the assumptions that $\eta_2(x, \theta_{2s})$ is continuous for all $\theta_{2s} \in \Omega_2$ and that the optimization problem (5.5.11) has a unique solution for ξ^* are sufficient to use all results from Chapter 3. If (5.5.11) does not have a unique solution, then the iterative procedure may not converge to an optimal design. This phenomenon is well-known in the general minimax optimization theory.

To assure the convergence of the iterative procedure to a "nearly" optimal design it is sufficient to make use of a regularization which is similar to what was discussed in Section 2.6: Replace ξ_s by $\overline{\xi}_s = (1 - \gamma)\xi_s + \gamma \overline{\xi}$, where $\overline{\xi}$ is a regular design, i.e., the least squares estimation problem

$$\hat{\theta}_2(\overline{\xi}) = \arg \min_{\theta_2 \in \Omega_2} \int [\eta(x) - \eta_2(x, \theta_2)]^2 \overline{\xi}(dx)$$

has a unique solution.

Example: Two Linear Models. As an example we consider a constant versus a quadratic model. The response functions are

$$\eta_1(x, \theta_1) = \theta_{11}$$
$$\eta_2(x, \theta_2) = \theta_{21} + \theta_{22}x + \theta_{23}x^2$$

with $|x| \le 1$. Suppose that the first model is true.

Without restrictions on the parameters, the adequacy of the constant model implies that of the quadratic model. Then, it does not make sense to perform experiments for detecting departures from the quadratic model because the noncentrality parameter will be identically zero. This situation is typical for so-called "nested" models where one model is a special case of the other model.

To make the models really different, we add the constraint $\theta_{22}^2 + \theta_{23}^2 \ge 1$. If the first model is true the non-centrality parameter is

$$\Delta_{21}(\xi) = \min_{\theta_{22}^2 + \theta_{23}^2 \ge 1} \int \left(\theta_{11} - \theta_{21} - \theta_{22}x - \theta_{23}x^2\right)^2 \xi(dx). \qquad (5.5.12)$$

The minimum will evidently occur on the set $\theta_{22}^2 + \theta_{23}^2 = 1$.

Let us proceed in a way that is typical for experimental design applications. First, we select some design that is intuitively reasonable, and then we check the optimality of this design using *Theorem 5.5.1*.

The symmetry of the design space and the constrained model η_2 imply the symmetry of the optimal design with respect to the origin. Therefore, we try the design

$$\xi^* = \left\{ \begin{array}{ccc} -1 & 0 & 1 \\ \frac{1}{4} & \frac{1}{2} & \frac{1}{4} \end{array} \right\}$$

which is "good" for the estimation of θ_{23} (see Section 2.5). Noting that two solutions of (5.5.12) exist for this design, viz.,

$$\theta_{21}^{(1)} = \theta_{11} - \frac{1}{2}, \quad \theta_{12}^{(1)} = 0, \quad \theta_{23}^{(1)} = 1,$$

and

$$\theta_{21}^{(2)} = \theta_{11} + \frac{1}{2}, \quad \theta_{22}^{(2)} = 0, \quad \theta_{23}^{(2)} = -1;$$

we state the sensitivity function to be

$$\phi_{21}(x, \xi^*) = \sum_{\ell=1}^{2} \zeta_\ell \left(\theta_{11} - \theta_{21}^{(\ell)} - \theta_{22}^{(\ell)} x - \theta_{23}^{(\ell)} x^2 \right)^2.$$

For any ζ_1 and ζ_2 such that $\zeta_1 + \zeta_2 = 1$, ϕ_{21} turns out to be $\phi_{21}(x, \xi^*) = (\frac{1}{2} - x^2)^2$ and has maximal values of $1/4$ at the points of the design ξ^* in question. This design, therefore, is T-optimal. Note that it does not depend on θ_{11}.

This independence of the optimal design from the parameters of the true model is, however, not the case if the second model is true. Without loss of generality we may assume that $\theta_{22}, \theta_{23} < 0$. Then the T-optimal design

$$\xi^* = \left\{ \begin{array}{cc} 1 & \max(-1, -\theta_{22}/2\theta_{23}) \\ \frac{1}{2} & \frac{1}{2} \end{array} \right\}$$

allocates observations at two design points where the discrepancies between the possible values of the second model is largest. The T-optimality of the design is an obvious corollary of *Theorem 5.5.1*.

Equivalent Design Problems. In (5.5.8) we have assumed that the first model is true. We also have assumed that the corresponding regression parameters are given. In the following, we abandon these two assumptions and consider the optimization problem

$$\xi^* = \arg\max_{\xi} \min_{\theta_1 \in \Omega_1, \theta_2 \in \Omega_2} \int [\eta_1(x, \theta_1) - \eta_2(x, \theta_2)]^2 \xi(dx). \qquad (5.5.13)$$

To simplfy notation, we introduce the function $\eta(x, \theta) = \eta_1(x, \theta_1) - \eta_2(x, \theta_2)$ and the vector $\theta^T = (\theta_1^T, \theta_2^T)$. Then, we may replace (5.5.13) by the simpler problem

$$\xi^* = \arg\max_{\xi} \min_{\theta \in \Omega} \int \eta^2(x, \theta) \, \xi(dx). \qquad (5.5.14)$$

If

$$\Psi(\xi, \theta) = \int \eta^2(x, \theta)\, \xi(dx) \qquad (5.5.15)$$

and $\eta(x, \theta)$ is a continuous function of x in X for all $\theta \in \Omega$, then we can state similar to *Theorem 5.5.1* that a necessary and sufficient condition for a design ξ^* to be T-optimal is the existence of a measure ζ^* such that for all $x \in X$

$$\phi(x, \xi^*) \le \min_{\theta \in \Omega} \int \eta^2(x, \theta)\, \xi^*(dx), \qquad (5.5.16)$$

where $\phi(x, \xi) = \int_{\Omega(\xi)} \eta^2(x, \theta)\, \zeta(d\theta)$, $\int_{\Omega(\xi)} \zeta(d\theta) = 1$ and

$$\Omega(\xi) = \left\{ \theta: \ \theta(\xi) = \arg\min_{\theta \in \Omega} \int \eta^2(x, \theta)\, \xi(dx) \right\}.$$

Note that in stating this optimality condition we have not assumed that the function $\eta(x, \theta)$ is linear with respect to θ. Thus, the above result has a very general character. To explore links between (5.5.14) and the optimization problems for models with linear response functions that are considered in Chapters 2 to 4, let us assume that $\eta(x, \theta) = \theta^T f(x)$. From definition (5.5.15) follows immediately that

$$\Psi(\xi, \theta) = \int \theta^T f(x) f^T(x) \theta\, \xi(dx) = \theta^T M(\xi) \theta. \qquad (5.5.17)$$

This expression for $\Psi(\xi, \theta)$ allows us to state the following results.

Theorem 5.5.2 *The design problem (5.5.14) is equivalent to the minimization of*

1. $c^T M^-(\xi) c$, if $\Omega = \left\{ \theta: (c^T\theta) \ge \delta > 0 \right\}$;

2. $\max_{x \in X} f^T(x) M^{-1}(\xi) f(x)$, if $\Omega = \{\theta: \eta^2(x, \theta) \ge \delta > 0, x \in X\}$;

3. $\lambda_{\max}\left(M^{-1}(\xi)\right)$, if $\Omega = \left\{ \theta: \theta^T\theta \ge \delta > 0 \right\}$.

Proof: 1. This result can be verified by direct minimization of the Lagrangian function $\theta^T M\theta + \lambda(c^T\theta - \delta)$. The minimum is attained at $\theta^* = M^- c / c^T M^- c$, and it is equal to $(c^T M^- c)^{-1}$. The latter proves the result. Note that regularity of the information matrix is not required.

2. Minimization of $\Psi(\xi, \theta)$ for $\eta^2(x, \theta) = (\theta^T f(x))^2 \ge \delta$ gives

$$
\min_{\theta:\max_x (\theta^T f(x))^2 \ge \delta} \theta^T M\theta = \min_x \min_{\theta:(\theta^T f(x))^2 \ge \delta} \theta^T M\theta
$$
$$
= \min_x \left(f^T(x) M^{-1} f(x) \right)^{-1}.
$$

Unlike in the previous case, the information matrix must be regular to assure that $f^T(x) M^{-1} f(x) < \infty$ for any $x \in X$.

3. The result follows from

$$\lambda_{\min}(M) = \min_\theta \frac{\theta^T M\theta}{\theta^T\theta}.$$

In this proof we have not made use of any specific property of ξ. This implies that *Theorem 5.5.2* is valid not only for continuous T-optimal designs but also for its exact versions. If we confine ourself to continuous designs, then the second part of the theorem may be extended. From *Theorem 2.4.1* follows the equivalence of the following optimization problems:

$$\max_{\xi} \min_{\theta : \eta^2(x,\theta) \geq \delta} \int \eta^2(x,\theta)\,\xi(dx)\,,$$

$$\min_{\xi} \max_{x} f^T(x) M^{-1}(\xi) f(x)\,,$$

$$\max_{\xi} |M(\xi)|\,.$$

Note that the equivalence of $\max_{\xi} |M(\xi)|$ with the above problems implies that D-optimal designs are also "good" for model discrimination.

D-optimal designs may also work well for another class of the testing problems. Let us assume that some *a priori* information on the parameters θ is available in form of the distribution function $\pi_0(d\theta)$. Then it is reasonable to use the mean of the noncentrality parameter as a criterion of optimality:

$$\Psi_0(\xi) = \int\int \eta^2(x,\theta)\xi(dx)\pi_0(d\theta)\,. \tag{5.5.18}$$

If the dispersion matrix of the *a priori* distribution π_0 is D_0, then we obtain for the case of the linear response function $\eta(x,\theta) = \theta^T f(x)$ that

$$\Psi_0(\xi) = \operatorname{tr} D_0 M(\xi)\,.$$

If D_0 is unknown, we may use the criterion

$$\Psi(\xi) = \min_{|D_0| \geq d} \operatorname{tr} D_0 M(\xi) = \min_{|D_0| \geq d} \Psi_0(\xi) \tag{5.5.19}$$

with some suitable $d > 0$; this criterion is based on a weaker constraint. If the matrix $M(\xi)$ in nonsingular we can use the inequality

$$\min_{|A| \geq d} \operatorname{tr} AB = m d^{1/m} |B|^{1/m}$$

[see relation (E) in the Appendix A] and obtain

$$\Psi(\xi) = m\, d^{1/m} |M(\xi)|^{1/m}\,. \tag{5.5.20}$$

The maximization of $\Psi(\xi)$ from (5.5.20) is equivalent to the maximization of $|M(\xi)|$. In other words, criterion (5.5.19) is equivalent to the D-criterion. Thus, we established another equivalence result:

Theorem 5.5.3 *The D-criterion and the criterion $\Psi(\xi)$ from (5.5.19) are equivalent.*

This result obviously holds both for the exact and for the continuous design problem.

The Tchebysheff Approximation Problem. Let us assume that $\eta^2(x, \theta)$ is a convex function of θ for any $x \in X$ and that Ω is compact, i.e., for any normed measure $\zeta(d\theta)$,

$$\eta^2(x, \overline{\theta}) \leq \int \eta^2(x, \theta)\, \zeta(d\theta), \tag{5.5.21}$$

where $\overline{\theta} = \int \theta\, \zeta(d\theta)$. Using (5.5.16) and (5.5.21), we can show that

$$\begin{aligned}
\min_{\theta \in \Omega} \max_{x \in X} \eta^2(x, \theta) &\leq \max_{x \in X} \eta^2(x, \overline{\theta}^*) \\
&\leq \max_{x \in X} \int \eta^2(x, \theta)\, \zeta^*(d\theta) \leq \min_{\theta \in \Omega} \int \eta^2(x, \theta)\, \xi^*(dx) \\
&\leq \min_{\theta \in \Omega} \max_{x \in X^*} \eta^2(x, \theta),
\end{aligned} \tag{5.5.22}$$

where $\overline{\theta}^* = \int \theta\, \zeta^*(d\theta)$, ζ^* is defined in the context of (5.5.16), and $X^* = \mathrm{supp}\, \xi^*$. Note that the second line in (5.5.22) is the restated inequality (5.5.16). However, we get

$$\min_{\theta \in \Omega} \max_{x \in X} \eta^2(x, \theta) \geq \min_{\theta \in \Omega} \max_{x \in X^*} \eta^2(x, \theta) \tag{5.5.23}$$

because $X^* \subset X$. Obviously, (5.5.22) and (5.5.23) do not contradict each other only if

$$\min_{\theta \in \Omega} \max_{x \in X} \eta^2(x, \theta) = \min_{\theta \in \Omega} \int \eta^2(x, \theta)\, \xi^*(dx). \tag{5.5.24}$$

From (5.5.24) we can infer the following result.

Theorem 5.5.4 *If $\eta^2(x, \theta)$ is a convex function of θ for any $x \in X$ and Ω is compact, then support points of a T-optimal design coincide with the extremal basis X^* of the Tchebysheff approximation problem*

$$(\theta^*, X^*) = \arg \min_{\theta \in \Omega} \max_{x \in X} |\eta(x, \theta)|. \tag{5.5.25}$$

If $\eta(x, \theta) = \theta^T f(x)$, the convexity of $\eta^2(x, \theta)$ is obvious.

Example: Test for Polynomial Regression. We want to test the polynomial of degree m versus the $(m-1)$-st degree polynomial. Let X be $[-1, 1]$. For a fixed coefficient of the term x^m we have to find

$$\xi^* = \max_{\xi} \min_{\theta} \int \left(x^m - \theta^T f(x)\right)^2 \xi(dx), \tag{5.5.26}$$

where $f(x) = (1, x, \ldots, x^{m-1})$. According to *Theorem 5.5.3*, the support set of ξ^* is defined as

$$X^* = \arg \min_{\theta \in \Omega} \max_{|x| \leq 1} |x^m - \theta^T f(x)|. \tag{5.5.27}$$

The optimization problem (5.5.27) is well known [see, for instance, Karlin and Studden (1966)]; for the support set we get

$$X^* = \left\{ x_i^* = \cos \frac{m+1-i}{m} \, \pi, \, i = 1, \ldots, m+1 \right\}.$$

The weights may be calculated directly and turn out to be $p_i = 1/m$ for $1 < i \le m$ and $p_1 = p_{m+1} = 1/2m$.

5.6 Nonlinear Regression

Locally Optimal Designs. The most specific feature of the optimal design problem for nonlinear response functions is the dependence of the information matrix $\underline{M}(\xi_N, \theta)$ upon the unknown parameters. Let us start with the very "unrealistic" case that θ is given. Similar to (2.3.6), we can define a locally optimal design

$$\xi^*(\theta) = \arg \min_\xi \, \Psi[M(\xi, \theta)], \qquad (5.6.1)$$

where ([cf. (1.4.9)]

$$M(\xi, \theta) = \int f(x, \theta) f^T(x, \theta) \, \xi(dx). \qquad (5.6.2)$$

Nothing is new in (5.6.1) as compared to (2.3.6), and all results from Chapters 2 and 3 may be applied without any changes. In particular, we may use the numerical methods of Chapters 3 to construct an optimal design $\xi^*(\theta)$ for a given θ.

Is this of any use for practical applications? The answer is positive if computations are not expensive (as it is nowadays the case) and if some *a priori* knowledge about θ is available. For instance, it may be known that $\theta \in \Omega$, and Ω can be covered with a moderate grid. By analyzing the behavior of

$$\Psi^*(\theta', \theta) = \Psi \left[M(\xi^*(\theta'), \theta) \right] \qquad (5.6.3)$$

for various values θ', we can find a design that is reasonably "good" for any θ from Ω.

This approach, in spite of its naivety, is very useful in practice. If the reader strives for more sophisticated results, then the minimax and Bayesian approach offer appropriate opportunities.

Minimax Design. The comparison of the information matrices defined by (2.6.24) and (5.6.2) reveals that they are identical except for a detail in the notation (u must be replaced by θ). This means that the results from Section 2.6 [as it is announced in the comment to (2.6.24)] may be used after obvious changes in notation. For instance, the minimax version of the D-criterion is

$$\Psi(\xi) = \max_{\theta \in \Omega} |M^{-1}(\xi, \theta)|. \qquad (5.6.4)$$

We may state (cf. Section 2.6) that a necessary and sufficient condition for ξ^* to be optimal in the sense of a (5.6.4) is the existence of a measure ζ^* such that

$$\max_x d(x, \xi^*, \zeta^*) \leq m, \tag{5.6.5}$$

where

$$d(x, \xi, \zeta) = \int_{\Omega(\xi)} f^T(x, \theta) M^{-1}(\xi, \theta) f(x, \theta) \, \zeta(d\theta) \tag{5.6.6}$$

and

$$\Omega(\xi) = \left\{ \theta : \; \theta(\xi) = \arg \min_{\theta \in \Omega} |M(\xi, \theta)| \right\}.$$

Note that the function

$$f^T(x, \hat{\theta}) M^{-1}(\xi, \hat{\theta}) f(x, \hat{\theta})$$

is the approximate variance of $\eta(x, \hat{\theta})$ at point x, where $\hat{\theta}$ is the l.s.e. of θ. From Carathéodory's Theorem follows that a measure ζ^* exists with not more than $m + 1$ support points.

Average or Bayesian Criterion. Similar to Section 2.6, we may introduce

$$\Psi_F(\xi) = \int_\Omega \Psi[M(\xi, \theta)] F(d\theta), \tag{5.6.7}$$

where $F(d\theta)$ represents the *a priori* knowledge on θ, in order to avoid the dependence of the optimality criterion on unknown parameters; (5.6.7) gives an average of $\Psi[M(\xi, \theta)]$ over θ.

Of course, $\Psi_F(\xi)$ is convex with respect to ξ if $\Psi(M)$ is convex with respect to M. Assuming that all assumptions from *Theorem 2.3.2* hold (if necessary, uniformly with respect to θ) for every θ from Ω, we may derive an analogue of this theorem. For instance, if $\Psi(M) = -\log|M|$, then a necessary and sufficient condition for ξ^* to be optimal is the inequality

$$\max_x d_F(x, \xi^*) \leq m, \tag{5.6.8}$$

where

$$d_F(x, \xi) = \int_\Omega f(x, \theta) M^{-1}(\theta, \xi) f(x, \theta) F(d\theta). \tag{5.6.9}$$

Unlike $\Psi(\xi)$ in Section 2.6, $\Psi_F(\xi)$ depends upon different matrices $M(\xi, \theta)$ for different θ. Therefore, Carathéodory's Theorem cannot be applied as it was done in the proof of *Theorem 2.3.1*, and the statement about the existence of an optimal design with not more than $m(m+1)/2$ support points is not generally valid in the present situation.

To construct an optimal design based on (5.6.7) we can use an analogue of the first order algorithm in Section 3.1; it is obtained by replacing $d(x, \xi_s)$ by $d_F(x, \xi_s)$. However, this is practically sound only if $F(d\theta)$ is discrete and contains only a few support points.

Averaging like in (5.6.7) is only one of several possibilities to avoid the dependence on θ. For the D-criterion, e.g., it may be expedient to introduce

$$\Psi_{k,F}(\xi) = \frac{1}{k} \log \int |M(\theta,\xi)|^k F(d\theta)\,.$$

The corresponding sensitivity function is

$$\phi_{k,F}(\xi) = \frac{\int |M(\theta,\xi)|^k d(x,\xi|\theta) F(d\theta)}{\int |M(\theta,\xi)|^k F(d\theta)}\,,$$

where

$$d(x,\xi|\theta) = f(x,\theta) M^{-1}(\theta,\xi) f(x,\theta)\,.$$

5.7   Design in Functional Spaces

A basic assumption that is maintained in almost all parts of the book (with some occasional exceptions like in Section 5.1) is that the design region X is a subset of the k-dimensional Euclidean space. However, to derive of the main results, e.g., the equivalence theorems, no use is made of this assumption. It is just a matter of convenience for discussions and examples. In this section we show how the main ideas and results from convex design theory may be applied for design regions of more complicated structure, say, when X is a subset of some functional space.

The Model. To keep the presentation simple we consider the standard model (1.1.1), viz. $y = \theta^T f(x) + \varepsilon$, with

$$f(x) = \int_U g(u) x(u)\, du\,, \tag{5.7.1}$$

where $g(u)$ is an m-vector of continuous functions on the compact set U and $x(u)$ is a function that satisfies certain constraints which correspond to a given experimental situation. For instance, we may assume that

$$X = \{x : 0 \le x(u) \le 1,\ u \in U\}\,. \tag{5.7.2}$$

In physics, the functions $x(u)$ are often called "slit functions". The design problem is how to choose the functions $x(u)$ in order to estimate θ optimally with respect to a certain criterion.

Models of this type are used in spectrometry for analyzing and describing spectra of mixtures (cf. the example on spectroscopic analysis in the section Illustrating Examples of the Introduction). In that context, the vector θ represents the concentrations of the m components in a mixture, the m functions $g_i(u)$ are optical spectra of these substances, and u stands for the frequency or the wave length, i.e., $u \in U \subset R$; $x(u)$ defines ranges of frequencies or wave lengths where observations are to be taken.

Optimal Designs and Properties. We confine the discussion to the one-dimensional case where $U = [a,b]$. The reader might specify more complicated models and explore the corresponding design problems. A rather extensive discussion of the topic gives

Ermakov (1983), Chpt. 7. Note that for the following results only the linearity of (1.1.1) and (5.7.1) is essential.

By definition the vector f is element of $X_f \subset R^m$; the design problem may be considered as a particular case of the design problem for the multivariate, (with respect to both parameters and independent variables) linear model

$$y = \theta^T f + \varepsilon, \ f \in X_f. \tag{5.7.3}$$

To use the results from Chapters 2 and 3, we only need to verify the compactness of X_f. All other assumptions are either obviously fulfilled or related to the optimality criterion to be applied. The fact that the set X_f is a compact, convex set in R^m is well-known in the moment spaces theory. The proof of convexity if obvious. The proof of compactness is beyond the scope of the book and can be found in the literature on moment spaces theory; see, for instance, Karlin and Studden (1966), Chpt. VIII.1. For D-optimality, we know from *Theorems 2.3.2* and *2.4.1* that the support set of an optimal design must belong to the set where the maximum of $f^T D(\xi) f$ is achieved. This can only occur on the boundary of X_f. Therefore, we are interested in this boundary.

Let $\underline{x} \in X$ be a function $x(u)$ which assumes the value one on a finite set of intervals in U and zero elsewhere. The end-points of these intervals are called the nodes of $\underline{x}$. The number of separate nondegenerate intervals on which $\underline{x}(u) = 1$ is called the index $In(\underline{x})$ of $\underline{x}$; an interval the closure of which contains an end-point of U, is counted as $1/2$; an interval that coincides with U is counted as 0.

If the functions $g(u) = (g_1(u), \ldots, g_m(u))^T$ constitute a Tchebysheff system, then a necessary and sufficient condition that $f(\underline{x})$ belongs to the boundary of X_f is that

$$In(\underline{x}) \le \frac{m-1}{2}. \tag{5.7.4}$$

Moreover, every boundary point of X_f corresponds to a unique $\underline{x} \in X$ [cf. Karlin and Studden (1966) Chpt. VIII.1].

If c_i, $i = 1, \ldots, m-1$, are closures of intervals on which $\underline{x}(u) = 1$ we can map any $\underline{x}$ into a set

$$C = \{c : \ a \le c_1 \le \ldots \le c_{m-1} \le b\}.$$

Thus, in the case of this special structure of $\underline{x}$, every (slit) function or "design point" $\underline{x}$ can be represented by a $(m-1)$-dimensional space.

An Example. Let $g_1(u) \equiv 1$, $g_2(u) = u$, $U = [-1, 1]$; we further assume that the functions $x(u)$ satisfy (5.7.2) and are of the type $\underline{x}$. From (5.7.4) follows that $In(\underline{x}) = 1/2$, and therefore, any boundary point of X_f is defined by one of the following three types of the slit function:

$$\underline{x}_1(u) = 1, \ 1 \le u \le 1;$$

$$\underline{x}_2(u) = \begin{cases} 1, & -1 \le u \le c_1, \\ 0, & \text{otherwise}; \end{cases}$$

$$\underline{x}_3(u) = \begin{cases} 1, & c_2 \le u \le 1, \\ 0, & \text{otherwise}. \end{cases}$$

Given that we are interested in a D-optimal design. From the so implied symmetry requirement follows that an optimal design must be searched only among designs with $c_2 = -c_1 = c$ and $p_1 = p$, $p_2 = p_3 = (1-p)/2$. Noting that

$$f_1(x_1) = \int_{-1}^{1} du = 2, \quad f_1(x_2(c)) = \int_{-1}^{-c} du = 1 - c,$$

$$f_1(x_3(c)) = \int_{c}^{1} du = 1 - c,$$

$$f_2(x_1) = \int_{-1}^{1} u \, du = 0, \quad f_2(x_2(c)) = \int_{-1}^{-c} u \, du = (c^2 - 1)/2,$$

$$f_2(x_3(c)) = \int_{c}^{1} u \, du = (1 - c^2)/2,$$

we find that

$$M(\xi) = M(c,p) = \begin{pmatrix} p + (1-p)1 - c)^2 & 0 \\ 0 & (1-p)(1-c^2)^2/4 \end{pmatrix}.$$

Direct maximization of $|M(c,p)|$ with respect to c and p lead to $c^* = 2 - \sqrt{5} \simeq -0.236$ and $p^* = \frac{3-\sqrt{5}}{4} \simeq 0.191$.

Design in L_2. Let us assume that $g(u)$ and $x(u)$ are functions from the Hilbert space $L_2(U)$ defined on the compact set U. Moreover, let the functions $g(u) = (g_1(u), \ldots, g_m(u))^T$ be a subset of some complete orthonormal basis $\{g_i(u)\}_1^\infty$, in $L_2(U)$ so that

$$\int g_i(u)g_j(u) \, du = \delta_{ij}, \quad i,j = 1, 2, \ldots . \tag{5.7.5}$$

Let our design region be defined by

$$\int_U x^2(u) \, du \le 1. \tag{5.7.6}$$

Applying the Fourier expansion gives

$$x(u) = \sum_{i=1}^{\infty} b_i g_i(u), \quad b_i = \int_U x(u) g_i(u) \, du; \tag{5.7.7}$$

we may replace (5.7.6) by the constraints

$$\sum_{i=1}^{\infty} b_i^2 \le 1 \tag{5.7.8}$$

on the Fourier coefficients. Combining (5.7.1), (5.7.5), and (5.7.7) we obtain

$$f_i(x) = \int_U g_i(u) \sum_{j=1}^{\infty} b_j g_j(u) \, du = b_i .$$

Introducing the vector $b = (b_1, \ldots, b_m)^T$ allows to write the regression model (1.1.1) with (5.7.1) as

$$y = \theta^T b + \varepsilon, \tag{5.7.9}$$

where $b^T b \leq 1$. The latter inequality is an obvious corollary of (5.7.8).

Thus, we have transformed the design problem defined by (1.1.1), (5.7.1), and (5.7.6) into a design problem for the first order polynomial model with spherical design region. This problem has been discussed in detail in Section 2.5. There, we concluded that any design with support points coinciding with vertices of any regular polyhydron inscribed in the m-dimensional sphere $b^T b \leq 1$ and with (among those points) uniformly distributed weights is optimal with respect to any optimality criteria; the optimality of this design is invariant with respect to rotation. The designs that are easiest obtained in terms of numerical effort consist of m support points with equal weights m^{-1} that coincide with the vertices of a regular simplex. Another choice may be vertices of any inscribed cube and in particular the cube with the vertices

$$
\begin{array}{cccc}
(\pm 1, & 0, & \ldots, & 0), \\
(0, & \pm 1, & \ldots, & 0), \\
\cdots & \cdots & \cdots & \cdots \\
(0, & 0, & \ldots, & \pm 1);
\end{array}
$$

the corresponding optimal design is

$$
\xi^* = \{\pm g_i(u), 2m^{-1}\}_1^m .
$$

Appendix A

Some Results from Matrix Algebra

This Appendix provides a few results that usually cannot be found in standard text-books.

(A) If A and B are positive definite matrices of order m and α is a real number with $0 \leq \alpha_s \leq 1$ then

(A1)

$$|\alpha A + (1 - \alpha)B| \geq |A|^{\alpha}|B|^{1-\alpha};$$

the equality holds if $A = B$;

(A2)

$$[\alpha A + (1 - \alpha)B]^{-1} \leq \alpha A^{-1} + (1 - \alpha)B^{-1};$$

the equality holds if $A = B$.

(B) If A is a regular matrix of order m and B an $(m \times p)$-matrix, then

(B1)

$$|A + BB^T| = |A|\,|I + B^T A^{-1} B|,$$

(B2)

$$(A + BB^T)^{-1} = A^{-1} - A^{-1}B(I + B^T A^{-1} B)^{-1}B^T A^{-1}.$$

(C)

$$(A + B)^{-1} = A^{-1} - A^{-1}(A^{-1} + B^{-1})^{-1}B^{-1}$$

(D) If A_j is an arbitrary $(r \times s)$-matrix and B_j is a positive-definite $(s \times s)$-matrix $(j = 1, 2)$, then

$$[(1 - \alpha)A_1 + \alpha A_2][(1 - \alpha)B_1 + \alpha B_2]^{-1}[(1 - \alpha)A_1 + \alpha A_2]^T$$
$$\leq (1 - \alpha)A_1 B_1^{-1}A_1^T + \alpha A_2 B_2^{-1}A_2^T,$$

(E) If A and B are positive-definite matrices of order m, then

$$\operatorname{tr} AB \geq m|A|^{1/m}|B|^{1/m}.$$

Appendix B

List of Symbols

$y, \boldsymbol{y}$	response or independent or observed variable
η	response function
$x, \boldsymbol{x}$	independent or control or design or regressor variables, predictors
t	uncontrolled regressor variables
θ	parameters
m	number of parameters
Ω	parameter space
$f(x)$	vector of basis functions
ε	(response) error
σ^2	variance of the error
ξ_n	discrete design
$\xi, \xi(dx)$	continuous design
ξ^*	optimal design
Ξ	set of designs
x_i	support points
X	support region
r_i	number of observations at x_i
p_i	weight of observations at x_i
n	number of support points
N	total number of observations
$M, M(\xi)$	normalized information matrix
$\underline{M}$	nonnormalized information matrix
$D, \underline{D}$	dispersion matrix of estimators $\hat{\theta}$
$d(x)$	variance of predicted response $\hat{\eta}(x, \hat{\theta})$
$\Psi, \Psi(M)$	optimality criterion
$\phi(x, \xi)$	sensitivity function for design ξ at x
$\Phi(\xi)$	constraint for design ξ

References

Albert A. (1972). *Regression and the Moore-Penrose Pseudoinverse*. New York: Academic Press.

Atkinson A.C. and Donev A.N. (1992). *Optimum Experimental Design*. Oxford: Clarendon Press.

Atkinson A.C. and Fedorov V.V. (1988). "Optimum Design of Experiments", p. 107-114 in S. Kotz and N.I. Johnson, *Encyclopedia of Statistics. Supplemental*. New York: J. Wiley.

Bandemer H., Bellmann A., Jung W., Le Anh Son, Nagel S., Näther W., Pilz J., and Richter K. (1977). *Theorie and Anwendung der optimalen Versuchsplanung. I. Handbuch zur Theorie*. Berlin: Akademie-Verlag.

Box G.E.P. and Draper N. (1987). *Empirical Model Building and Responce Surfaces*. New York: J. Wiley.

Brown L.D., Olkin I., Sacks J., and Wynn H.P. (Eds.) (1985). *Jack Kiefer. Collected Papers III. Design of Experiments*. New York: Springer.

Cook R.D. and Fedorov V.V. (1995). "Constrained Optimization of experimental design", *Statistics*, **26**, 129-178.

Ermakov S.M. (1983). *Mathematical Theory of Experimental Design*. Moscow: Nauka.

Fedorov V.V. (1996). "Design of Spatial Experiments: Model fitting and prediction", pp. 515-553 in S. Ghosh and C.R. Rao (Eds.), *Handbook of Statistics, Vol.13*. New York: Elsevier Science.

Fedorov V.V. (1972). *Theory of Optimal Experiments*. New York: Academic Press.

Fedorov V.V. and Malyutov A.B. (1974). "Design of Experiments in Regression Problems", *Mathematische Operaationsforschung und Statistik*, **3**, 281-321.

Ghosh S. and Rao C.R. (Eds.) (1996). *Handbook of Statistics, Vol.13*. New York: Elsevier Science.

Kanwal R.P. (1971). *Linear Integral Equations, Theory and Techniques*. New York: Academic Press.

Karlin S. and Studden W. (1966). *Tchebysheff Systems: With Applications in Analysis and Statistics*. New York: J. Wiley.

Laurent P.J. (1972). *Approximation et Optimization*. Paris: Hermann.

Nalimov V.V., Golikova T.I., and Granovsky Y.V. (1985). "Experimental Designs in Russian Practice", pp. 475-496 in A.C. Atkinson and S.E. Fienberg (Eds.), *A Celebration of Statistics: The ISI Centenary Volume*. Berlin: Springer-Verlag.

Pázman A. (1986). *Foundations of Optimum Experimental Design*. Dordrecht: Reidel.

Pilz J. (1993). *Bayesian Estimation and Experimental Design in Linear Regression Models*. New York: J. Wiley.

Pshenichnyi B.N. (1971). *Necessary Conditions for an Extremum*. New York: Marcel Dekker.

Pukelsheim F. (1993). *Optimal Design of Experiments*. New York: J. Wiley.

Rao C.R. (1973). *Linear Statistical Inference and Its Applications, 2nd Edition*. New York: J. Wiley.

Ripley B.D. (1981). *Spatial Statistics*. New York: J. Wiley.

Seber G.A.F. and Wild C.J. (1989). *Nonlinear Regression*. New York: J. Wiley.

Silvey S.D. (1980). *Optimal Design*. London: Chapman and Hall.

Smith K. (1918). "On the Standard Deviations of Adjusted and Interpolated Values of an Observed Polynomial Function and its Constants and the Guidance they Give towards a Proper Choice of the Distribution of Observations", *Biometrika*, **12**, 1-85.

Szegö G. (1959). *Orthogonal Polynomials*. New York: American Mathematical Society.

Wynn H.P. (1984). "Jack Kiefer's contribution to experimental design." *Annals of Statistics*, **12**, 416-423.

Index

Lecture Notes in Statistics

For information about Volumes 1 to 50
please contact Springer-Verlag